T0010320

Pleasant Valley

Pleasant Valley

LOUIS BROMFIELD

with an Introduction by Gene Logsdon
and Drawings by Kate Lord

The Kent State University Press
KENT, OHIO

Published by
The Kent State University Press
Kent, Ohio 44242

Pleasant Valley originally published in 1945 by Harper Brothers.

Library of Congress Catalog Number 55006568
ISBN 978-1-60635-461-2
Published in the United States of America

Cataloging information for this title is available at the Library of Congress.

27 26 25 24 23 5 4 3 2 1

For Bob and Virginia, Max and Marian, Kenneth and Martha, Harry and Naomi, Pete Gunder, Wayne Hastings, Jack McGuggin, George Terris, Chuck Hubbs, Jimmy Caddick, Johnny Roodhuyzen, Dave Stamper, Dick McFalls and all the others who have had a part in the adventure of Malabar and most of all for Marjory who always thought it a good idea.

CONTENTS

PREFACE

THIS BOOK IS A PERSONAL TESTAMENT WRITTEN OUT OF A lifetime by a man who believes that agriculture is the keystone of our economic structure and that the wealth, welfare, prosperity and even the future freedom of this nation are based upon the soil.

The book is not written for agricultural experts, although they may find in it observations of interest, but for the average reader who does not know too much about the earth and what goes on in it and above it. I think it will be understood by the good husbandman who loves his soil and his animals and by the man or woman who finds that for himself there is nothing so exciting or so satisfying or so beautiful as the earth and the seasons and rich green fields and fat cattle, the sound of foxes barking in the night and the raccoon's print in the snow.

It is, frankly, a romantic book, written in the profound belief that farming is the most honorable of professions and unquestionably a romantic and inspiring one.

INTRODUCTION

WRITING AN INTRODUCTION TO LOUIS BROMFIELD'S *PLEASANT VALLEY* involves for me such an eerie coincidence as to be downright scary. For a person who says he does not believe in fate, I am good evidence that some things seem fated to happen. I read *Pleasant Valley* the first time in the mid-forties and then again in the mid-fifties when I was trying to find what home meant for me. My mother brought it home from the library in about 1945 when it was first published, shoved it into my hands, and commanded, "Read this!"

I was twelve years old and there were parts of the book I didn't much understand, but the parts that focused on farms and farm life I loved, just as my mother did.

Had someone told me that someday I'd write an introduction for that book I would have swooned away in a dead faint. The reason Mom thought the book important was that Bromfield knew how to make farming sound like the honorable profession that it was. "A good farmer is always one of the most intelligent and best educated men in the community," Bromfield wrote. And again; "We are apt to forget that the man who owns land and cherishes it and works it well is the source of our stability as a nation."

In our lives, we knew first hand the kind of prejudice and ridicule that urbanism liked to throw at farmers, so much so that many farmers themselves felt inferior. Not my mother. She considered farming the most noble of all careers, and now she had a book—*Pleasant Valley*—to back her up. Because of her and that book, I grew up feeling the same way.

But it was not until I reread the book recently, in preparation for doing this introduction, that I realized just how much of an intellectual debt I owed to Louie Bromfield. Reading now, I see, almost with embarrassment, how many of the statements I have made in print over the past 25 years sound so like his, in some instances so identical as to seem almost plagiaristic:

> *Neither Max* [his foreman] *nor myself believed that the impulse of our times toward regimentation, centralization, mechanization and industrialism necessarily represented progress.*

> *I was perpetually haunted by the terrible economic insecurity which a mechanized and industrial civilization imposes upon the individual.*

> *Someday I would like to compile a book . . . dedicated to the things that have been taught more for the benefit of manufactures of farm machinery, of chemical fertilizers, and of prepared and expensive feeds than for the good of the earth or the welfare of farmers.*

> *I had memories of the farm of my grandfather, a farm which in itself was a fortress of security.*

> *Working with Nature we would be recompensed by her sympathy but beyond a certain point Nature could not be pushed.*

> [The farmer] *is, I think, the happiest of men for he inhabits a world that is full of wonder and excitement over which he rules as a small god.*

That I would be asked to write an introduction to the re-issue of this book I loved as a child and that influenced me as much as any book as a young man, is fateful enough, but that is only half the story.

I have just finished a book of my own called *You Can Go Home Again*. *Pleasant Valley* is, if it is anything, an account of going home again successfully. Bromfield in the early forties had just escaped the German occupation of France where he and his family lived for more than a decade, and he begins *Pleasant Valley* with an account of arriving back home, fearful that the haunts of his childhood might all have been swallowed up by the monster of progress. To his delight, however, the real "Pleasant Valley" was still about the same as it was in the early '30s, just as I found my home grounds relatively unchanged in 1974, when I returned, from what it had been in 1960 when I left. Although the rest of *Pleasant Valley* covers a great variety of subjects, it essentially chronicles how a person, even of such insatiable curiosity and restlessness as Bromfield, could come home successfully, just as I did. Sorry about that, Thomas Wolfe.

> *A man could have that good kind of life whether he had fifty acres or five thousand.*

> *. . . whole fields of dogtooth violets and trillium and Canada lilies and Dutchman's breeches and bloodroot. And in Summer the same woods were waist high in ferns and snakeroot and wild grapes hung down from among the branches so that the whole woods seem a tropical place. . . .*

> *I have any number of friends who spent all their lives as bankers and industrialists or workingmen or insurance salesmen only to discover at middle age that in reality they were farmers all the time without knowing it.*

> *And I am grateful to Louis Gillet . . . for he sent me back to the county where I was born, to Pleasant Valley and the richest life I have ever known.*

The reason *Pleasant Valley* and other "agricultural" books by Bromfield have a permanence far beyond that of his many novels is that to those who know Ohio farms and farmers, his knowledge of them is unerringly genuine. In this day and age, it is so satisfying to read someone who knows what he is talking about, especially in the field of agriculture where really knowing the subject has always been viewed by publishers (who don't know either) as incidental. Very few good writers know how to farm well, and very few good farmers know how to write well. The chapter titled *"My Ninety Acres"* is so good, so genuine, so perfect that for years I believed that it was true, not fictional. In a higher sense it is true, truer than any non-fiction I've ever read. And because it was so true, it led me to establish a "my ninety acres" of my own.

Sometimes in articulating the foundations of conservation farming that were being laid down at that time, largely through his writing, Bromfield got carried away a little as any enthusiastic writer will do. He should be read as one reads about an explorer or researcher. Not even Columbus was one hundred percent right, nor for that matter Darwin. When Bromfield waxes eloquent about letting weeds grow in corn, he lets the emotion of one discovery cloud his overall judgement. Of course low weeds under the corn will help break the fall of raindrops and because of that (and other reasons) the soil in a weedy cornfield might hold more moisture and therefore might make a bigger corn crop some years. But I doubt if Bromfield would have remained enthusiastic about that idea if he had to run the cornpicker at harvest time, when those low weeds have a terrible tendency to wrap around the rollers and cause farmers to invent marvelous new cuss words. And, even growing under the corn, those weeds do go to seed. If he had thought a little more about the adaptability of hybrid corn, he might have foretold the future in this regard as well as he predicted it in others; eventually we would plant corn a lot thicker than farmers did in the forties, which would break the fall of raindrops better than low weeds and without necessarily perpetuating weed problems.

But in general, Bromfield's ecological farming instincts and practices (I believe he was the first American agriculturist to use the word *ecology* frequently) were unerringly correct and they remain so today. He viewed the farm as an organic whole: the garden as important as the cornfields; the woodlots as important as the hayfields; the wild animals as important as the livestock; the people as important as the financial balance sheet. If a civilization does not learn to do likewise, it is doomed.

Indeed today, Bromfield can be read to better purpose by gardeners, homesteaders, and part-time, small-scale farmers than by commercial agriculture, where, at least for the time, big factory methods and money are the only things that count. "An acre is as good an anchor as five thousand," says Bromfield, "for in that acre, as Fabre well knew, there is a whole universe and the answer to most of man's problems." Amen.

But if ever there was a case where an author's personal character ought to be considered apart from his farm writing, Bromfield is the one. He comes close to being two totally different people: Bromfield the contemplative farmer/writer and Bromfield the bombastic socialite; or as I like to put it, Bromfield, the careful planter of crops in rotation, and Bromfield the profligate sower of wild oats in exhilaration. And yet the two personae are integral parts of his whole. Every writer worth his salt must have a keen sense of the dramatic. If you can't find drama, even in a field of corn (or especially in a field of corn), readers are not going to read you no matter what. That is a fact of life. But that sense of the dramatic, that flair for a good story, is both the writer's glory and if not controlled, the writer's downfall. The ability to "write a good story" moves by very slow increments, often unnoticed by the writer himself, to becoming an entertainer outside his writing. Sooner or later, the writer is then tempted to turn himself into a spectacle for public consumption.

Bromfield gave in to this temptation. He became a showman and his farm a showplace. That he knew showmanship was not essential to what he was doing is indicated when he introduces the chapter on the grossly huge house he built on Malabar with this note: "A chapter to be skipped by those who have no interest in architecture."

It should surprise no one that he entertained the high and mighty of Hollywood and politics at Malabar Farm. Instinctively he understood that part of him was like them, a showman. An actor. The only problem is that being a showman is destructive to being a writer. Writers need to be invisible so that they can observe the world more accurately and deeply.

Actually Bromfield had little choice. His readers demanded showmanship of him. They came trooping to Malabar by the thousands on weekends, wanting to be entertained. His publishers loved it, since showmanship sells books. (It is so much worse today, when publishers insist that writers rush from bookstore to bookstore, making spectacles of themselves.) Furthermore, he was intent upon a most impossible mission, trying to convince the powerful monetary forces of the Agribusiness Army that profiteering in farming was wrong. Only by being the most consummate showman could he ever hope to make any impression at all on that front, going against the grain of short term profit. My favorite story about Bromfield is told by his long and faithful foreman on the farm, Max Drake (from *Return To Pleasant Valley*, edited by George DeVault (THE AMERICAN BOTANIST, 1996):

> *. . . At the banquet that night* [in Des Moines, Iowa] *I could see that Louie was set, primed because he had that group of people* [in front of him]. *Now mind you, here were some top-flight best farmers in Iowa, the best research, I mean everybody who knew anything about agriculture was there that night. . . . And he had them eating out of his hand. In his talk not only were the banks going down that river, the outhouses were going down the river, the farms were going down the river and he laid it on. "Now," he says, "what you've got to do is what we have done at Malabar! We started to do contour farming. The soil has responded. We've got springs that have started flowing that never flowed for years." Well, we cleaned 'em out, that's the reason they flowed. He didn't say that.* [He said it was] *because we were doing strip cropping. And, he said, "our oat yields were up 50 percent last year. And our corn yields"—every time he pulled a fact out of the air like that I'd go down in my seat, I was sitting right up beside of him. When it was over, believe me, they all lined up to shake his hand. . . . I got off in the corner and was kind of enjoying what was going on and somebody put their arm up over my shoulder . . .* [and said], *"Max, don't worry about it. Somebody's got to scare the be-jesus out of 'em." I never worried about it again. I don't care how much he exaggerated, he got his points across. He got people concerned and just was great. I loved the guy.*

Me too.

—GENE LOGSDON

Pleasant Valley

THE RETURN OF THE NATIVE

As the car came down out of the hills and turned off the Pinhook Road, the whole of the valley, covered in snow, lay spread out before us with the ice-blue creek wandering through it between the two high sandstone ridges where the trees, black and bare, rose against the winter sky. And suddenly I knew where I was. I had come home!

All the afternoon we had been wandering through the southern part of the county trying to find the Pleasant Valley Road. It was not as if I had never been there before. Once I had known it very well, as only a small boy can know a valley where the fishing and swimming is good, the woods are thick and cool and damp, and filled with Indian caves. I had known every turn of the creek, every fishing hole, every farm, every millrace, every cave. But that had been a long time ago—more than thirty years—and now finding my way back through the hills was like trying to find one's way back through the maze of a vaguely remembered dream. There were places I remembered when I came upon them, places like the village of Lucas, and the bridge over the Rocky Fork and the little crossroads oddly called Pinhook and another called Steam Corners, no one has ever been able to say why.

But like scenes and places in a dream these were isolated landmarks, disconnected, with the roads that lay between only half-forgotten mysteries. In those hills where the winding roads trail in and out among the woods and valleys, I can lose myself even today after I have come to know the country all over again as a grown man. On that first day I was utterly lost.

I might have asked my way. In the village there was the little bank of which one day I was to become a director, and the village post office where one day I was to do what I could, as an amateur politician, to get Hoyt Leiter appointed postmaster. As I write this I am struck again by the curious dreamlike quality of the whole adventure in which the elements of time and even of space seemed confused and even suspended. It was as if the Valley had been destined always to be a fiercely dominant part of my existence, especially on its spiritual and emotional side. It has always existed for me in two manifestations, partly in a dreamlike fashion, partly on a plane of hard reality and struggle. Perhaps these two manifestations represent the sum total of a satisfactory life. I do not know. I think that some day when I am an old man, it, like many other mysteries, will become clear, and that clarity, as one of the recompenses of old age, will become a part of the pattern of a satisfactory life.

I could have stopped and asked my way at the bank or the post office, or I could have stopped the car and asked one of the three or four people who passed us trudging along on foot in the snow. Once we passed a boy in a Mackinaw and ear-muffs riding on a big blue-roan Percheron mare. He raised his hand in greeting. As I passed each one I thought, "That might be a Teeter or a Berry or a Shrack or a Darling or a Tucker or a Culler."

Those were all names which had belonged among these hills since Indian times. The young ones I couldn't know because all of them had been born during the thirty years I had been away. But I peered into the faces of the old ones trying to find there something that I remembered from the days when, as a small boy, I had driven over the whole of the county in a buggy behind a team of horses, electioneering for my father or for some other good Democrat.

Then, as a small boy, I had known all the Teeters, the Cullers, the Berrys and all the others for sometimes we had had midday dinner with them. Sometimes we had tied the horses to the hitching rail and got out and gone into the fields to help ring hogs or husk corn. Sometimes, if the roads were deep with mud or an early blizzard came on, we unharnessed the steaming horses and put them in a stall deep with straw and ourselves spent the night in a big bed with a feather quilt over us. I think my father was welcome in the house of any farmer or villager of the county, Republican or Democrat. They all knew him as "Charley" Bromfield. He was a kindly man, who was a bad politician because he didn't pretend to like people; he really liked them. I think he liked politics, in which he never had much success and in which he lost a great deal of money, principally because it brought him into close contact with nearly all kinds of people. He was one of the fortunate people who liked the human race despite all its follies and failings.

And so the faces of the people I saw in the village streets and on the country roads were very important to me on that first day. I was coming home to a country which I had never really left, for in all those years away from the Valley it had kept returning to me. It was the only place in the world for which I had ever been homesick. More than half of the time I had spent away from Pleasant Valley was passed in France, a country where I had never been a stranger, even on that first night when I stepped ashore at Brest at the age of eighteen and tasted my first French cheese and French wine in the smoky, smelly little bars and cafés of the waterfront. Often in distant parts of the world, among strange peoples, I had wakened to find that I had been dreaming of Pleasant Valley.

There have been moments in my experience when I have been sharply aware of the "strange intimations" of which Dr. Alexis Carrel writes—intimations which have scarcely been touched upon in the realms of science—"strange intimations" of worlds which I had known before, of places which in the spirit I had touched and heard and smelled. France was one of the places I had always known. From the time I was old enough to read, France had a reality for me, the one place in all the world I felt a fierce compulsion to see. Its history fascinated me, its pictures, its landscapes, its books, its theaters. It was, during all my childhood and early youth, the very apotheosis of all that was romantic and beautiful. And finally when, the morning before we were allowed ashore, the gray landscape of Brittany appeared on the horizon, there was nothing strange about it. I had seen those shores before, when I do not know. And afterward during all the years I lived there, during the war when I served with the French army and in the strange, melodramatic truce between wars, it was always the same. Nothing ever surprised or astonished me; no landscape, no forest, no château, no Paris street, no provincial town ever seemed strange. I had seen it all before. It was always a country and its people a people whom I knew well and intimately.

I have had a similar feeling about the austere, baroque, shabby-elegant quality of Spain and about the subcontinent of India. Germany, under any regime, has always been abhorrent, a place where I was always depressed and unhappy and a hopeless foreigner, even in a city like Munich which many people accept as beautiful and warm and *gemütlich*—a feeling which was not improved by spending my last visit there in the Vierjahreszeiten Hotel with Dr. Goebbels. And although, save for a little very remote Swedish blood, I have no blood that did not come from the British Isles, England was always a strange, although very pleasant country, more exotic to me than Spain or India. I do not begin to understand these things, these "strange intimations."

The point I wish to make is this—that during all those thirty years, sometimes in the discomfort of war, sometimes during feelings of depression engendered by Germany, but just as often during the warm, conscious pleasure and satisfaction of France or India or the Spanish Pyrenees, I dreamed constantly of my home country, of my grandfather's farm, of Pleasant Valley. Waking slowly from a nap on a warm summer afternoon or dozing before an open fire in the ancient presbytère at Senlis, I would find myself returning to the county, going back again to the mint-scented pastures of Pleasant Valley or the orchards of my grandfather's farm. It was as if all the while my spirit were tugging to return there, as if I was under a compulsion. And those dreams were associated with a sensation of warmth and security and satisfaction that was almost physical.

It may have been because in all my waking hours, during most of those years, I was aware of insecurity and peril, conscious always that in the world outside my own country, a doom lay ahead. During the last few years before the end of Europe, the feeling of frontiers, hostility and peril became increasingly acute, and distant Pleasant Valley, fertile and remote and secure, seemed more and more a haven, hidden away among the lovely hills of Ohio. I think no intelligent American, no foreign correspondent, living abroad during those years between the wars, wholly escaped the European sickness, a malady compounded of anxiety and dread, difficult to define, tinctured by the knowledge that some horrifying experience lay inevitably ahead for all the human race. Toward the end the malady became an almost tangible thing, which you could touch and feel. Many doctors had a hand in the attempt to check it, most of them quacks like the Lavals and the Daladiers, some untrustworthy like Sir John Simon, later honored by a title for his follies and deceptions, some merely old-fashioned practitioners with quaint nostrums patched and brewed together like Neville Chamberlain. One saw the marks of the malady on every face, from the hysterical ones whose follies became even more exaggerated, to the dull or unscrupulous ones who became each an individual thinking only of himself, and rarely of his country.

Toward the end I found myself spending more and more of my sleeping hours in the country where I was born and always what I dreamed of was Ohio and my own county.

And at last when Mr. Chamberlain debased the dignity of the British Empire, took his umbrella and overshoes and went to Munich to meet a second-rate adventurer, I, like any other moderately informed and intelligent person in Europe, knew that the dreadful thing was at hand, and that nothing now could stop it. I sent my wife and children home to America on a ship crowded with schoolteachers and businessmen and tourists whose pleasure trips or business had been cut short by orders from the State Department for all Americans traveling in Europe to come home.

My wife said, “Where shall we go?” and I replied, “To Ohio. That is where we were going anyway sometime.”

I myself stayed on, partly out of a novelist’s morbid interest in the spectacle, however depressing, and partly because, loving France, I wanted to be of help if there was anything that I could do. I stayed for weeks, more and more depressed, and it was Louis Gillet who persuaded me at last that I could do more for France at home in my own country than I could ever do by remaining in France.

I remember that we talked it all out beneath the great trees and among the magnificence of the ruins of the Abbaye de Chaalis where Louis Gillet held the sinecure of curator. The leaves were falling from the trees in the long gentle autumn of the Ile de France. His four boys were all mobilized. His widowed younger daughter was there on a visit from her quiet farm in Périgord. A second daughter, wife of the head of the Institute de France in Athens, and her two small children, had not returned to Athens at the end of the summer because of the malady, of that doubt and dread which crippled the will and the plans of everyone in Europe, in homes in Poland, in Norway, in Italy, in England.

And as we walked about the great park among the lagoons and the seventeenth century gardens, surrounded by the evidence of all the glorious history of France, Louis Gillet talked, brilliantly, humanly as he could talk when he was deeply moved.

At last he said, "You must return home. There is nothing you can do here that a Frenchman could not do. You can go home and tell your people what is happening here, what is bound to come. Tell them they will not be able to escape it—to be prepared and ready. We in France and in England too have already lost half the battle by complacency and bitterness and intrigue. The Hun is preparing to march again down across the face of civilized Europe. Go home and tell your people. You can help France most by doing just that."

That night after dinner all of us went out into the moonlit forest of Ermenonville to listen to the stags call. It was the mating season and the stags made a wonderful roaring noise to attract the does. Sometimes in a patch of moonlight you could, if you were down wind and sat quite still, catch a glimpse of a big stag calling, his head raised, his muzzle thrust straight out, sick with love. And if you were very lucky you could witness the magnificent spectacle of two stags fighting over a doe. The deer were the descendants of those deer which François Premier and Henry Quatre had hunted in these same forests. It was the fashion in the autumn on moonlit nights to go out from the provincial towns in the Oise to listen to their calling—*entendre bramer les cerfs.*

It had been the fashion perhaps as far back as the days when our town of Senlis was a Roman city.

Sitting there in the warm sand and lanes of the forest on a moonlit night, surrounded by a family which represented all that was finest in France and therefore in our Western civilization, I experienced a faint sickness in the pit of my stomach. In a day or two I would be leaving all this—the forest, the old town of Senlis, the good people who lived there. I would be saying farewell to France which I had loved and known even before I had ever seen it. And if one day I returned it would never be the same. It would live, because an idea, a civilization never wholly dies but goes on living in some altered form as a contribution to all that follows, but it would be changed, dimmed and dissipated by the violence of war and decadence. I would never again find the France I was leaving.

I was aware too, quite suddenly, of what it was that attracted me to Europe and most of all to France; it was the sense of continuity and the permanence of small but eternal things, of the incredible resistance and resiliency of the small people. I had found there a continuity which had always been oddly lacking in American life save in remote corners of the country like parts of New England and the South which were afflicted by decadence, where permanence and continuity of life existed through inertia and defeat. In the true sense, they were the least American of any of the parts of America. They had stood still while the endless pattern of change repeated itself elsewhere in factories, in automobiles, in radio, in the restlessness of the rich and the nomadic quality of the poor.

The permanence, the continuity of France was not born of weariness and economic defeat, but was a living thing, anchored to the soil, to the very earth itself. Any French peasant, any French workingman with his little plot of ground and his modest home and wages, which by American standards were small, had more permanence, more solidity, more security, than the American workingman or white-collar. worker who received, according to French standards, fabulous wages, who rented the home he lived in and was perpetually in debt for his car, his radio, his washing machine.

Sitting there it occurred to me that the high standard of living in America was an illusion based upon credit and the installment plan, which threw a man and his family into the street and on public relief the moment his factory closed and he lost his job. It seemed to me that real continuity, real love of one's country, real permanence had to do not with mechanical inventions and high wages but with the earth and man's love of the soil upon which he lived.

I knew that the hardest thing for me to bear in leaving France and Europe was not the loss of the intellectual life I had known there, nor the curious special freedom which a foreigner knows in a country he loves, nor the good food, nor even the friends I would be leaving behind. The thing I should miss most, the thing to which I was most attached were the old house and the few acres of land spread along the banks of a little river called the Nonette—land, earth in which I had worked for fifteen years, planting and cultivating until the tiny landscape itself had changed. If I never saw it again a part of my heart would always be there in the earth, the old walls, the trees and vines I had planted, in the friendships that piece of earth had brought me with horticulturists, farmers, peasants, market gardeners and the workingmen whose communal gardens adjoined my own.

They had liked and respected me, not because I was by their standards fabulously rich or because on Sundays Rolls-Royces and automobiles labeled *Corps Diplomatique* stood before my door. They liked and respected me because I grew as good or better cabbages with my own hands than they were able to grow. And it occurred to me that the honors I valued most out of all those I had received was the diploma given me by the Workingmen-Gardeners' Association of France for my skill as a gardener and the medal given me by the Ministry of Agriculture for introducing American vegetables into popular cultivation in the market garden area surrounding the city of Paris.

All of these things had to do with a permanence, a continuity which one seldom found in America. When I returned home, I knew that permanence, continuity, alone was what I wanted, not the glittering life of New York and Washington, not the intellectual life of universities. What I wanted was a piece of land which I could love passionately, which I could spend the rest of my life in cultivating, cherishing and improving, which I might leave together, perhaps, with my own feeling for it, to my children who might in time leave it to their children, a piece of land upon which I might leave the mark of my character, my ingenuity, my intelligence, my sense of beauty—perhaps the only real immortality man can have so the people would say long after I was dead, as they would say in Senlis long after I was gone, "Yes, the American did that. He planted that tree and built that bridge. He made the garden below the river in the old orchard." I cannot see that man could wish a better afterlife than the peace of oblivion and the immortality that rests in houses and trees and vines and old walls.

But on the floor of the forest the November fog had begun to settle down and the first chill of winter had begun to slip in about us. The stags, satisfied, had quit calling. Quietly we walked back to the Abbaye under a waning moon, past the canals built six hundred years before by the monks of Chaalis who were also the first millers of that rich wheat country.

Inside the pleasant house under the pompous bourgeois portrait of the Gillet ancestor who had been a marshal under Louis Philippe, we had a glass of good vin rosé and I drove home at last through the forest back to Senlis. It was the last time I ever saw Louis Gillet, with his long, sallow, bearded face and blue eyes. Afterward, week after week, I had letters from him when he was a *réfugié*, proscribed by the Germans, living in Montpellier with one son gravely wounded, another a prisoner in Germany and a third with the Free French in Syria. He kept on fighting with his pen and voice against the Nazis, against German *kultur*, against the defeatism and treason of Vichy. And then we entered the war and the Germans occupied all of France, a steel curtain came down, and I heard from him no more.

By some curious chance on the day I began to write this book, I had word through the French Underground that Louis Gillet was dead, away from the Abbaye which he loved, a *réfugié* in Montpellier. I do not know whether the Germans imprisoned him, or how he died, but I do know now, six years after we walked beneath the trees in the Abbaye gardens, that all the advice he gave me was wise advice, that all the things he feared and predicted have come true. I know that he was a wise and good man. My only regret is that he never came to see me in Pleasant Valley, to see the land to which I had returned because we both believed it was my destiny and a good destiny. I would have liked him to know, to understand through his own senses, the rightness of all he said that November evening in the forest of Ermenonville before what must have been for him—the end of the world.

All along, all through the years of homesickness and even after I had come back to America, I had never said to anyone that the county in which I was born was one of the beautiful spots of the earth. I had kept the belief to myself, a little out of shyness, a little because there were times when I, myself, had doubts, knowing that all too often when later in life you revisit scenes you have known and loved as a child, something strange has happened to them. Somehow, mysteriously as you grew into manhood and swallowed the whole of the world, they have become shrunken and different. The houses that you remembered as big and beautiful have dwindled and become commonplace, the stream on which you once played pirates is no longer a lovely gleaming river but has turned into a small and muddy brook.

On that winter afternoon while I searched for Pleasant Valley among the hills and winding roads, I was a little afraid that when I came suddenly upon it, I would find that it had changed, that all the while I had been dreaming of something that no longer had any existence in reality.

With me in the car were Mary, my wife, and George who had been my friend and managed my affairs for a great many years. You will hear of them again in the course of this story, so it is just as well to explain them now. Mary was born on Murray Hill in New York City, a New Yorker of New Yorkers, and George was born on Long Island. Neither had ever known Ohio—my Ohio—save for the flat uninteresting country south of the lakes where Mary had taken the children on her flight from the doomed maneuvers of Chamberlain and Hitler.

All that afternoon as the car drove southward out of the flat lake country into the rolling hills of Wayne and Ashland County they kept saying, "This is beautiful country, why didn't you tell us about it?"

I had never really talked about it but once. Years earlier during a fit of homesickness in Switzerland, I had written a whole book about it called *The Farm*. I knew now that until the car turned into the wooded hill country, they hadn't really believed what I had written then. They had thought the county and the people I described were imagined as in a book of fiction. Their exclamations encouraged me; perhaps, after all, the Valley would be exactly as I had remembered it.

And then we turned the corner from the Pinhook Road and I knew that I was right. Nothing had changed. It lay there in the deep snow, wide and pleasant between the two high sandstone ridges covered by forest. Halfway up the slopes on each side, in the shelter of the high ridges, stood the familiar houses and the great barns, unchanged after thirty years—houses with the old names of the Pennsylvania Dutch and old English stock which had settled the country long ago—the Shrack place, the Mengert place, the Berry place, the big white houses and barns of the Darling settlement set in the wide flat rich end of the Valley where Switzer's Run joined the Clear Fork.

And then far away, a mile or more on the opposite side of the Valley I saw a small house with an enormous cupolaed barn. The buildings sat

on a kind of shelf halfway up the long sloping hill that turned its back on the north winds. It was already twilight and the lower Valley was the ice-blue color of a shadowed winter landscape at dusk and the black, bare trees on the ridge tops were tinted with the last pink light of the winter sunset. There were already lozenges of light in the windows of the distant house. Like Brigham Young on the sight of the vast valley of Great Salt Lake, I thought, "This is the place."

I heard my wife saying, "What a lovely, friendly valley!"

On that late winter afternoon, one had a curious sense of being sheltered from the winter winds, from the snow, from the buffetings and storms of the outside world. My wife and George saw a snow-covered valley. They could not see what I was seeing for the Valley had no place in their memories. What I saw was a spring stream in summer, flowing through pastures of bluegrass and white clover and bordered by willows. Here and there in the meanderings of the stream there were deep holes where in the clear water you could see the shiners and bluegills, the sunfish and the big red-horse suckers and now and then a fine small-mouthed bass. On a hot day you could strip off your clothes and slip into one of these deep holes and lie there in the cool water among the bluegills and crawfish, letting the cool water pour over you while the minnows nibbled at your toes. And when you climbed out to dry in the hot sun and dress yourself, you trampled on mint and its cool fragrance scented all the warm air about you.

I saw, too, fields with fat cattle and wild marshy land where the cattails grew ten feet high and the muskrats built their shaggy round houses in the autumn, marshes which in April were bordered and splashed with the gold of one of the loveliest of all wild flowers, the marsh marigold. A little later in summer from among the rich tropical green of its spade-shaped foliage the arrowroot threw up long spikes of azure blue. And I saw the old mills, high, unpainted, silver-gray with the weathering of a hundred years, the big lofts smelling of wheat and corn and outside the churning millrace where fat, big carp and suckers lay in the deep water to feed on the spilled grain and mash.

And I saw not the winter-naked woods, all snow and ledges of pink sandstone rock, but whole fields of dogtooth violets and trillium and Canada lilies and Dutchman's-breeches and bloodroot. And in summer the.same woods were waist high in ferns and snakeroot and wild grapes hung down from among the branches so that the whole woods seemed a tropical place like Brazil and Sumatra. As a boy in these woods I had pretended that they were tropical forests and that I was lost in them, as very often I was. And now I knew I was right. I had been far in the years between. I had seen tropical forests in Malabar and Macassar which held the same feeling of dampness, of fertility where, as in these Ohio woods, the leaves and tendrils and fresh green shoots were so thick that the whole air seemed green as if one were under water.

And I saw the woods in late February and March when there were no leaves on the naked trees and here and there in damp hollows the first lush green of the skunk cabbage was thrusting through the dead leaves and marsh grass of the year before. For me there is always something exciting and especially beautiful about the skunk cabbage. Boldly it thrusts its tropical green leaves into the frosty air of dying winter, the first of all plants to herald the awakening and rebirth of life with each spring.

That time of bareness when the skunk cabbage first appeared was the time of making maple syrup and I saw the sugar camp with its roaring fires and the woods streaked with the last melting snow and the fat horses steaming as they drew the sled which carried the great hogshead of fresh sap taken from the budding trees. And the long nights when the run was good and I was allowed to sit up all night with my grandfather and boil syrup while the fire made shadows on the oaken walls of the sugar camp and the wind howled outside.

I was seeing all this which the two others could not see. My heart was crying out, "Wait until spring comes! If you think it is nice now, you will see something you cannot even imagine when this country awakens."

We crossed the Valley and the little river half frozen over, with the swift-running clear spring water fringed with ice and rime, and up the hill on the opposite side of the wide sheltered ledge where the small house sat with its little windows blocked with light.

We followed the Hastings Road, a narrow, insignificant township road which led back and forth through woods and up and down low hills to the casual crossroad settlement called Hastings; and halfway up the hill we turned across a ravine with a small spring stream flowing down it, showing blue where the living water, green with cress, ran clear of ice between the dead leaves of last year's sweet flag.

At the house no one answered the knock. I knew it was chore time and so I went to the big barn to find the owners.

It was a big, red barn built in the days when farmers were rich and took a pride in their barns. Ohio is filled with them, barns which are an expression of everything that is good in farming, barns in which their owners took great pride. Nowadays one sees often enough great new barns on dairy farms owned by great corporations, or stock farms owned by millionaires; but these new barns have no character. They express nothing but utility and mechanized equipment, with no soul, no beauty, no individuality. Already they appear on any country landscape commonplace and standardized without beauty or individuality—in fifty years they will simply be eyesores.

The old barns built in the time of the great tradition of American agriculture when the new land was still rich and unravaged by greed and bad farming, had each one its own character, its special beauty born of the same order of spirit and devotion which built the great cathedrals of Chartres or Rheims or Salzburg. They were built out of love and pride in the earth, each with a little element of triumphal boastfulness—as if each barn was saying to all the rich neighboring countryside, "Look at me! What a fine splendid thing I am, built by a loving master, sheltering fat cattle and big-uddered cows and great bins of grain! Look at me! A temple raised to plenty and to the beauty of the earth! A temple of abundance and good living!"

And they were not built *en série*, like barracks. Each rich farmer had his own ideas, bizarre sometimes, fanciful with fretwork and cupolas and big handsome paintings of a Belgian stallion or a shorthorn bull, the main cupola bearing a pair of trotting horses bright with gilt as a weather vane. They were barns with great, cavernous mows filled with clover hay, two stories or three in height with the cattle and horses below bedded in winter in clean straw, halfway to their fat bellies. Perhaps there was waste space or they were inconveniently planned for doing the chores, but there was a splendor and nobility about them which no modern hip-roofed, standardized, monstrosity can approach. Ohio is filled with them—Gothic barns, Pennsylvania Dutch barns with stone pillars, New England barns attached to the house itself, the stone-ended barns of Virginia and even baroque barns. There is in Ohio no regional pattern of architecture as there is in New England or the Pennsylvania Dutch country. Ohio was settled by people from all the coastal states each bringing his own tradition with him, and so there is immense variety.

In my boyhood nearly all these barns had a rich, well-painted appearance. Those owned by farmers with an ancient Moravian background outdid the barns which only had a single stallion or bull painted on them; they had painted on the big sliding barn door a whole farm landscape for which the farm itself had served as a model and in it appeared bulls and cows, calves and stallions, hens and ducks and guinea fowl, horses and sheep and hogs. They were hex-paintings and their roots lay, not in Ohio or even in the coastal states, but far back in the darkness of medieval Germany, in a world of Bald Mountains and *Walpurgisnächte*. They were painted there on the big barn doors as a safeguard against the spells of witches, against vampires and incubi for it was believed and it is still believed among the old people that the spell cast by any malicious neighboring witch on the cattle in one of these great barns would fall not on the cattle themselves but upon the representations painted on the barn door. Always they were painted artlessly by someone on the farm and some of them had a fine primitive quality of directness and simplicity of conception.

Usually over the doors of these painted barns there hung a worn horseshoe, for it was believed that witches had an overweening passion for mathematics coupled with a devouring curiosity. If a witch sought during the night to sweep through the barn door on her broomstick and found herself confronted by a used horseshoe, she was forced to turn about and have no peace until she had retraced and counted all the hoofprints made by the shoe. The more worn the shoe the better, for it would take her all the longer to satisfy her compulsion, and she would not have completed her impossible task before morning arrived and she had to return whence she came. If the shoe had been worn long enough, the prints it had made would be so numerous that she could never count them all in a single night. As each night she had to begin afresh, she would never be able even in the long nights of winter, to get through the door to do evil to the cattle.

As a boy I had seen in the early mornings little heaps of corn or corn meal outside each door of a barn owned by some old man whose Moravian blood took him far back into the mists and shadows of Germany. They were placed before the doors for the same reason as the omnipresent horseshoe. A witch confronted by a heap of corn could not go on with her evil purpose until she had satisfied her curiosity by counting every grain. If the corn were ground into meal, so much the better for the task became a thousand times more difficult.

All these memories came flooding back during the short walk from the house to the great barn. Then I pushed open the door and walked into the smell of cattle and horses and hay and silage and I knew that I had come home and that never again would I be long separated from that smell because it meant security and stability and because in the end, after years of excitement and wandering and adventure, it had reclaimed me. It was in the blood and could not be denied. But all of that story I told long ago in *The Farm*.

UP FERGUSON WAY

Once the three farms we bought had all belonged to a man named John Ferguson. He had come to the Valley at the end of the eighteenth century when the Valley Road was no more than a trail frequented by Indians moving north or south to and from Lake Erie and the Ohio River. There are no more Fergusons in the Valley and so there are no very clear records concerning who John Ferguson was or whence he came. Certainly John Ferguson was Scottish and very likely he came into the Valley on foot for at that time there was not even a wagon trail and all the country was covered by almost impenetrable forest. Where he is buried I do not know nor does anyone else; perhaps among the trees of the virgin forest which he loved, in a grave long since disappeared. The graves of his descendants lie in the little graveyard beside a grove of oaks and hickory above the bottom pasture on the Fleming place, close by the graves of my father and my sister. Whether he brought a wife and children with him or not, no one knows any longer, but he left children, for their bodies sleep in the graveyard and their names appear in the huge pile of deeds that went with the three farms. In the beginning he had been a hunter and a trapper but in 1817 he became the owner of 640 acres of virgin land by virtue of a deed signed by James Monroe, fifth President of the United States.

From that date onward the recorded deeds represent a tangled history of the original 640 acres being divided and redivided, reunited and divided again by inheritance, by sale, by trading. There is the record of the parcels of land sold off to the two millers, Rose and Talbot by name, good names for millers, and once there had been a distillery with a still pond of the clear soft spring water which made excellent Bourbon whisky. By 1890 there were no more Fergusons for the name came to an end with a spinster and two bachelor brothers who lived on the part of the original 640 acres that lay high up against the sky with a view over three counties. That part of the farm is still known as the "Ferguson place."

In the pile of deeds lay all the history of that land, of marriages, and births and deaths, of quarrels and bargaining and bankruptcies, of the strange sterile end of the Ferguson blood in the little house by the spring on the top of the farm, of the highwaymen who terrorized the Valley as late as the beginning of this century, and the story of "Ceely" Rose, the miller's daughter who murdered her mother and father and two brothers.

And now all that land has been brought together again in a single piece after a hundred and thirty years by someone who had been drawn back to the Valley as John Ferguson had been drawn to it when he gave up his wandering life as hunter and trapper to settle there.

We bought three farms—the Fleming place, the Anson place and the Ferguson place. There remained outside only a pie-shaped slice, mostly marsh and wilderness, belonging to a bank and known as the Jungle. No one quite knows who gave it the name but it could not have a better one. It was all marsh and forest, filled with springs, with the little river running through it. In midsummer it becomes very nearly impenetrable. In it live muskrat and mink, raccoon and possum and a wonderful assortment of birds. Pheasants and quail find cover among the blackberries, the elderberries and the sumac which border it.

It wasn't only that the history of those 640 acres was written in the great pile of yellowing deeds and legal documents; it was written too in the very earth and trees and buildings, a sad history of rich land slipping downhill over a period of more than a century. The history was sadder even than the history of the births and deaths, the crimes and the murders written into the deeds. It was the story of good earth being murdered by carelessness and bad farming and greed and ignorance.

The worst of the three farms and the most beautiful was the Ferguson place. I do not know why John Ferguson chose it as the site of his original cabin or why his descendants clung to it until at last the line died out with two lonely old bachelors and an old maid. Given the choice, a shrewd farmer would never have selected it; he would have taken the Anson place which lay halfway down the gentle slopes of the Valley, or the Fleming place which spread across the rich fields in the Valley bottom. My only guess is that John Ferguson wasn't really a farmer at heart but a hunter and a trapper who loved not the rich orderly green of the cornfield but the wildness of the marsh and forest where there were deer and bear and foxes, and the fierceness of the elements—the blizzards which swept down the Valley in winter and the wild thunderstorms of the hot Ohio summer. And I think he must have loved solitude, for solitude and remoteness were and still are the very essence of the "Ferguson place."

It lies high up against the sky, a hill farm surrounded by woods and rocks and a ravine so deep that when the rest of the land was stripped of its trees, the virgin forest was left there because man had not yet devised a way of getting out the great logs. You approach the Ferguson place by a steep, half-ruined road through the forest, a road worn deep, where even the outcropping sandstone is rutted with ancient wheel tracks. On its damp shaded banks grow ferns and carpets of the loveliest of all spring flowers—the hepatica anemone. The road, even today, is half ruined because it is steep and all the year round the water which drains from the bank seeps into it, undoing the work done on it the year before. It is as if Nature herself sought to keep the Ferguson place wild and lonely and unviolated. Overhead the trees join their branches so that the whole road is a tunnel laced with wild grapevines where the light itself is the color of watery depths.

At the very top of the ruined road you come out suddenly into a sheltered pasture of bluegrass, surrounded by woods and dotted with black walnut trees. Above you is only the sky and the summit of the hill which rises yet higher above the pasture. The half-ruined road climbs another hundred feet and in a little grove of sassafras you come out on the top of the world. Immediately below you there spread out the pastures and fields of the Ferguson place and beyond, stretching into misty blue infinity lie the valleys and the rolling wooded hills of three counties

What you find in the view from that hilltop is not the wild, rugged beauty of mountain country which terrifies the soul and reduces the ego to insignificance. Nor do you have the feeling of imprisonment which high mountains bring to some people.

It is a pleasant land all about you, valleys where the bottom land is rich, bordered by hills covered with wild and luxuriant forest, the whole filigreed with the silver of the streams called Switzer's Run, Possum Run and the Clear Fork; and far down lies the blue shield of Pleasant Hill Lake bordered by the deep red of sandstone bluffs and the blue black of hemlock trees. Below in the rich land bordering the lake stand the big prosperous white barns and houses of the Darling settlement, and above on the shelving hills lie the smaller farms, their fields tucked away safely and neatly among the woods. Even today they have a look of having been carved out of the forest itself. On a brilliant day in May or October the whole sky above is filled with vast white clouds casting deep blue shadows that drift across the valleys and change the deep green of the hemlock to black and the fresh green of the new wheat fields to deep emerald.

On the top of the Ferguson place you have solitude in the midst of beauty and plenty. There on the hilltop you feel remote from cities and strife and mankind itself; yet close at hand down at the far end of the damp green tunnel you come out again into the world.

There is something *fey* about the Ferguson place, something which undoubtedly attracted John Ferguson and kept his descendants there, enchanted through four generations until at last his seed died out and the Ferguson place was left to become a wilderness once more. The legend in the Valley is that the Fergusons were always a little "teched." I think I understand why. It is a spot one turns to instinctively when all the world seems collapsing over one's head. You turn to it when fear and depression assail you, with that instinct for returning to the womb which Freud believed was strong in all of us. In the lofty wilderness and solitude of the Ferguson place one goes back to the beginning of time.

It is not myself alone who feels all this about that lonely hill farm. All the children and the grownups on the farm share the feeling. At times when I have gone there with the dogs to escape troubles and worries and like Antaëus to regain my strength by touching earth once more, I have discovered suddenly that I am not alone, that far down the slopes of the bluegrass pasture there is another solitary figure, dwarfed by the immensity of sky and earth. It is someone else come up from one of the houses on the farm below to get away from the burden of living and find strength and courage once more. And always each of these solitary figures pretends not to see the other and slips away like a shadow to lose himself in the woods. Always in the course of the day the children find their way to the Ferguson place as if something special drew them there. On all that lonely poor hill farm only one field is cultivated today; the rest has been allowed to return to healing pasture and woods. But that is the favorite field of everyone who works on the farm. No one minds working there alone for he returns at night stronger and fresher than when he climbed up to the lonely farm in the morning.

On the far side of the summit of the hill, you descend until you come to the ruins of an old orchard, and a hole in the ground where once there had been a house. There are grapevines climbing over a ruined arbor and here and there in the grass a clump of asparagus or rhubarb gone wild. The house was burned long ago, after the two bachelors and the old maid died, no one knows how, perhaps by a careless spreading fire left behind by hunters; but the ghostly touch of man's hand is still there in the big lilacs, the huge flaming japonica bush, the violets that hide in the shady corners and the drift of star-of-Bethlehem blossoms that has spread far beyond the borders of the old garden out into the orchard and the pasture.

Before the blackened stones of the doorstep stand two great Norway spruce, the mark of every old farmhouse throughout our county. Their sides are blackened by the scars of the fire and long since healed. Up one of them climbs a trumpet vine which in September fills the dark, weeping, branches of the spruce with the flame of its blossoms. And along the bank by the side of the ruined lane grow great clumps of the old-fashioned red day lilies which every settler's wife seemed to have brought with her from the East because they were so easy to grow. The lilies found pleasure in the Ohio country and like the old-fashioned single red rose, have gone wild everywhere, along roadside arid fences and even in woods.

The only building left on the Ferguson place is the trappers' cabin built long ago at the end of the eighteenth century by John Ferguson. It stands beside the spring, slowly rotting away, sinking back into the ground whence it came. Each year time and moisture and worms rot the big hand-hewn logs a little more. The spring beside it is a kind of miracle for it exists not at the foot of a hill in the Valley but almost on the top of a hill. Springs happen like that in our county, in unexpected and unlikely and unreasonable places, because underground lie vast strata of sandstone, porous and cracked by the movement of the earth and the weight of the glaciers which pushed their vast weight through the county a million years ago. Whenever the cracked and creviced stone comes near the surface there is a spring.

A little below the old orchard and the ruined log house, the land breaks away suddenly into a deep ravine filled with virgin forest where on the hottest, driest day of summer, spring water trickles over the rocks at the bottom and the air is green and moist and cool. At the head of the ravine the little brook trickles over the face of a great cave weathered out of the sandstone rock. All the ledges and the great overhanging shelf of rock drip moisture and are covered with ferns and wild columbine. Beneath, inside the damp cool cave, one goes back to the beginnings of time, for in the damp ooze of decaying sandstone grow primitive forms of algae and ferns that have survived there somehow from the days when all the world was a steaming swamp. In summer the cave is damp and cold and in winter the frost rarely penetrates its full depth. In the days when Indians moved north and south in the Valley, they stopped here and in the big cave on the other side of the Ferguson place, halfway down the hill not far from the ruined roadway. Both caves had shelter and spring water.

The other cave is of a different formation, much bigger and deeper and higher, a great narrow fissure created when the sandstone cracked and fell apart beneath the weight of the glacier. It is so well hidden, even today, that you can pass within a few feet of its mouth without discovering it. Now and then the children come home with an arrow or spearhead dropped there long ago by some wandering Delaware or Miami.

Last summer while digging for a pond below the big cave, we came upon an Indian grave. It was a simple affair. They had dug a pit in the hillside and filled it with gravel to keep it dry. Under the earth, side by side, lay the skeletons of six warriors killed in battle. The arm and shoulder of one of them had been shattered and the skull of another smashed by a tomahawk.

That is the Ferguson place, a beautiful, wild and haunted farm. In the Valley since the beginning there had always been a phrase which the neighbors used in referring to it. When they spoke of it they called it, "Up Ferguson Way" as if they were talking of something remote and different. Everyone on the farm speaks of it that way today, quite naturally. To go there is like going into another world.

But behind the beauty and wilderness there lay tragedy—the haunting tragedy of rich farm land ruined and scarred with gullies. When we came there only two fields were still worth cultivating. The rest had gone back into shaggy half-wild hill pasture with a seedling forest stealing in softly, imperceptibly, to heal over again the scars of man's evil and greedy treatment.

Below the Ferguson place, halfway down the side of the long Valley slopes stood the Anson place. You came upon it as you emerged from the high green tunnel of the ruined road, its house and big barns on a wide terrace shelf sheltered from the north by a high cliff of red and pink sandstone. It was a farm altogether different from the Ferguson place, lying warm and snug and compact facing the south and east so that it had the sun all day long until evening when it fell into the long blue shadows of the sandstone cliffs behind it.

The early settlers had a knack of picking out the best sites on which to raise their tiny cabins. The man who chose this spot had loved the land rather than the wild lonely forest which John Ferguson chose, and he had picked a sheltered spot. He had chosen a good spring bubbling out of the hillside and over it he built his cabin. In the early days this was a common practice for Indians roamed all the countryside, at first warlike and hostile, and at last as the white man crowded them out, drunken, disorderly and treacherous. Sometimes they lingered for days in the thick forest near a lonely cabin and clearing waiting to murder a man or carry off a child on the way from the house to the spring for water. In those days it was a good thing to have the spring inside the house.

The cabin has long since disappeared but the high foundations and the wall that enclosed the spring are still there, the big stones crudely chiseled out of the soft sandstone from the ledge above. The old stones are now part of the garden, overgrown with moss and ferns and wild columbine. There is a special story about the spring, of its death and rebirth, but that will come later in this book.

The worst massacres in the Valley were committed not in the days when warriors roamed the countryside but a little later when the Indians degenerated into dirty, drunken animals. Here and there among our hills, modest stone monuments have been raised on bloody ground where whole families were murdered in some lonely cabin, and the legend of Johnny Appleseed hangs over the whole Valley for this was his country and more than once he appeared in the middle of the night to warn some lonely cabin that marauding Indians were on the way.

JOHNNY APPLESEED AND AUNT MATTIE

My earliest memories of Johnny Appleseed are of listening to my Great-Aunt Mattie talk of him beneath the big catalpa tree on my grandfather's farm. Aunt Mattie was blind from the age of thirty and when I first remember her she was over eighty, a sprightly, very bright old lady with a crinkly mouth that was always curling up in a good-humored, faintly mocking smile. She was a witty, and at times a malicious, old lady and like so many blind people since the time of Homer, a great storyteller. I think that the stories were a kind of compensation for the darkness in which she spent more than half a century of her life. Now, nearly forty years after her death, I realize that you could not always take all her stories as gospel truth, but I also know that she was in her way a minor artist. If there were facts missing from one of her stories of frontier life, she supplied them out of her own imagination; if some fact did not suit one of her tales she modified and altered it to fit the artistic frame.

Aunt Mattie said she had known Johnny Appleseed. I do not know whether this was true or not. She was born in 1826 and by a curious combination of circumstances her presence and her stories brought me as a small boy very near to the eighteenth century for she was the child of my great-grandfather who had visited Voltaire at Ferney. She was born when he was seventy-two years old. But there were other elements which brought her very close to the strange little man who has become a legend and a kind of saint in our Middle Western country. She loved the woods and the streams and the wild birds and animals as Johnny Appleseed had done. My grandmother told me that when Aunt Mattie was a small child she had caused much anxiety through her trick of running off to spend whole days wandering through the swamps and forests of the still half-conquered Ohio frontier country. In those days there were occasional bears or wandering Indians about but no amount of bloodcurdling tales ever succeeded in instilling fears in Aunt Mattie as to what might happen to her. My grandmother said that, like Johnny himself, Aunt Mattie never seemed to have any fear of Indians or wild animals.

Even as a blind old woman she kept that love of the streams and forest and wild things. She had a remarkable talent for finding her way about the farm. Of course, she had been born there and until she lost her sight she knew it well and so even in all the fifty or more years of her blindness she must have known always exactly how it looked. But sometimes on long excursions she did not go off by herself, feeling with her small feet the roads and paths or guiding herself by the sound of the rustling leaves of the familiar landmark trees. Sometimes when she felt adventurous and wanted to make a long excursion down through the bottom pastures where the creek ran on into the thicket, she would ask one of the children to act as her guide. She would select a spot which she remembered beneath a tree on the edge of the creek and then tell us to come back for her two or three hours later. She did not like us playing about. She wanted to be left alone and at times, even as a child, you had the feeling that she had come there for a rendezvous and did not wish to be disturbed or spied upon. She would spend a whole afternoon listening to the sounds of fish jumping or birds singing or cattle lowing.

Since she knew that whole small world through touch and sound alone she undoubtedly understood it in a way none of the rest of us could ever do. She heard and interpreted sounds, small sounds—the symphony made by frogs and crickets and birds and cattle which we never heard at all. Sometimes you would come upon her sitting quietly beneath a tree beside the clear little stream, her head tilted a little in an attitude of listening. Like as not, the cattle would be lying all around close up to her, munching bluegrass and fighting flies.

The spirit of Johnny Appleseed haunted that same Valley. Once, long ago, he had roamed all the region, sleeping in the big sandstone caves or in Indian huts or settlers' cabins. He was welcome wherever he stopped among the Indians, the white settlers or the wild animals themselves.

With each year the figure of Johnny Appleseed grows a little more legendary; each year new stories and legends attach themselves to what has become in our country an almost mythical figure. A few facts are stated but few are known. It is said that Johnny's real name was John Chapman. Some say that he was born in New England, others that he was born at Fort Duquesne, later to be called Pittsburgh. It is pretty well accepted that he was a Swedenborgian by faith. It is also related that he died at last somewhere near the borders of Ohio and Indiana.

The truth is, of course, that Johnny Appleseed has attained that legendary status where facts are no longer of importance. Long before we returned to Pleasant Valley he had become a kind of frontier saint about whom had collected volumes of folklore and legend. In the natural process of things, it is the stories and legends and not the facts which have become important. I think Aunt Mattie understood this change of values which throughout all history has imperceptibly translated heroes into gods and hermits into saints.

She told us children that she remembered him well as an old man when he came to spend nights at her father's big farm. He was a small man, Aunt Mattie said, with a shriveled, weather-beaten face, framed by long ragged gray hair. His eyes were a very bright blue surrounded by fine little lines which came of living always in the open. He went barefoot winter and summer and for clothing wore strange garments fashioned out of a kind of sackcloth or of leather or skins given him by the Indians. His only baggage was a metal cooking pot with a handle, which did not encumber his movements since when traveling he wore it as a hat with the handle at the back. He always carried a "poke" swung over his shoulder in which he carried seeds and plants.

He would arrive in the evening and have supper with the family, although later on as he grew older and more solitary he would not eat in the house but only on the doorstep or in the woodshed. Sometimes in the evening he would preach a kind of sermon upon love of mankind and all Nature. As he grew older the Swedenborgian doctrines changed imperceptibly into a kind of pagan faith which ascribed spirits to trees and sticks and stones and regarded the animals and the birds as his friends. But the sermons never had the curse of the conventional doctrinal harangue; they were interspersed with wonderful, enchanting stories about the wild things, so that for the children, the opossum, the raccoon, the bear, the blue jay all came to have distinct personalities and a sense of reality which most people never understand. Aunt Mattie said that, like St. Francis, he had a habit of talking aloud to the birds and animals as he tramped bare-footed through the woods. None of the children ever resented or avoided his "sermons." It is probable that Johnny's visits took the place in that frontier country of theater and talking pictures and comic strips all rolled into one.

He never accepted the hospitality of a bed but chose instead to sleep in the great haymows above the fat cattle and horses. Usually when the settler went to the barn in the morning Johnny had already vanished with his kettle on his head and his "poke" of apple and fennel seed thrown over his shoulder. I think every Indian, every settler, every trader in all that Ohio country must have known him well, much as my great-grandfather knew him.

A good many of the white men and their families humored him and were fond of him but looked upon him as half mad. Some of them owed their lives and the lives of their families to Johnny Appleseed. The Indians regarded him with awe and veneration, for in the way of primitive peoples, they looked upon his particular kind of "insanity" as God-given, an "insanity" which linked him to the trees; the rocks, the wild animals which were so much a part of the redskins' daily and hourly existence. Because of this veneration the Indians never harmed him and left him free to go and come as he liked, nor did they conceal their plans from him. More than once during the early days of the frontier Johnny slipped away with his kettle and "poke" out of an Indian encampment to journey miles through forest and marsh to warn some lonely family to leave their cabin for the safety of the nearest village or blockhouse until the raids were over. If the Indians knew he had used his friendship to betray their plans, they appeared to have borne him no ill will for he lived at peace among them, preaching brotherhood and good will until there were no longer any Indians left in all the region and Johnny died one night, an old man, in a hedgerow in Indiana.

My Great-Aunt Mattie used to tell about Johnny's friendship with the King of France and although I do not remember that she ever claimed to have seen the King, she certainly intimated that she had a speaking acquaintance with him. She meant, of course, the Lost Dauphin, that curious and mysterious character who, as a young man, spent much time in our Ohio country. To the whites he was known as Eleazar Williams and to the French traders and the redskins he was known simply as Lazare. For many years he lived among the Indians and the French half-breed *coureurs de bois*, wandering down into our valleys where, undoubtedly, he met and knew Johnny. My great-aunt also said that when the king, Louis Philippe, then the Duc de Chartres, came into the Ohio country to investigate the claims and legend of the Lost Dauphin, he saw Johnny Appleseed and questioned him concerning his friend Eleazar Williams or Lazare. That too is not only possible but probable, for the Duc de Chartres, later King Louis Philippe, did come to Ohio accompanied by a commission to determine whether or not Eleazar Williams was really the son of Louis XVI and Marie Antoinette who had been spirited out of the Temple and off to America.

More than a hundred years later when I wrote of Johnny Appleseed and his friendship with the Lost Dauphin in *The Farm* and the book was translated into French, the ancient controversy was reopened in France and learned articles on the subject appeared in the *Revue des Deux Mondes*, *L'Illustration*, and other French periodicals, and the name and legend of Johnny Appleseed became known to countless Frenchmen of another time and world.

The whole tale of the Lost Dauphin is one of the fascinating stories of history and I think no one, even today, could say with any certainty that the Eleazar Williams who frequented our frontier country was or was not the son of Louis XVI and Marie Antoinette. His past, like that of Johnny Appleseed, has joined the legends and folklore of the Middle Western country. It is known that there arrived in Maine mysteriously at the end of the eighteenth century, a boy of ten or eleven years of age with a box of clothing. He spoke only French and appeared to be dim-witted or to have been frightened out of his wits by some terrifying experience. His only memories were those of mirrors and mobs and torches. A New England preacher named Williams took him in, it is not clear whether by adoption or as a kind of bound-boy. As he grew older the boy's wits never seemed quite normal and presently he slipped away to live on the frontier among Indians and French trappers. My Great-Aunt Mattie said he never spoke English very well and that he was blond and big and heavy with a big nose and a small chin. I doubt that Aunt Mattie knew what were the physical traits of the Bourbons but she described them perfectly in her account of Lazare, the Lost Dauphin. It is likely that she never saw him or was too young to remember him and that the description was passed on to her by her father who had visited at Ferney the old man who did more than any other to bring about the French Revolution. The Duc de Chartres, sent by the Dauphin's sister, (the dull, half-tragic Duchess d'Angoulême), to the wilderness of Ohio to investigate the stories, repudiated the Lost Dauphin, Lazare, but that was inevitable because the Dauphin stood between the Duke and the throne of France.

I like to think of crazy Johnny Appleseed and the Lost Dauphin with his dull wits, wandering our country, protected and respected by Indians because they were both "naturals" and thus close to the beasts and trees.

After I had written of Johnny and the Dauphin in *The Farm*, I had two letters, both from very old ladies, one in Illinois and one in Minnesota. The old lady from Illinois wrote that at one time there was in the state capital of Illinois an unmistakable portrait of Louis XVI which had been brought from France by a deputation of men who believed that Lazare was the son of Louis and Marie Antoinette and sought to persuade him to return to France as a pretender. My correspondent claimed that as a young girl she had seen the portrait but did not know what had become of it. From Minnesota another old lady wrote that she had known the false Dauphin as an old man married to an Indian squaw. Her description of him was, like that of Great-Aunt Mattie, unmistakably the description not only of a Bourbon but of a son of Louis XVI—that of a fat, soft, blond old man with a scant beard, almost more feminine than masculine. I doubt that anyone will ever know whether Lazare was the true or the false Dauphin. I only know that in legend he has gone down as a friend of our Johnny Appleseed, and that to Aunt Mattie he was always the true King of France.

There is some disagreement concerning the way in which Johnny went about planting apple trees in the wild frontier country. Some say that he scattered the seeds as he went along the edges of marshes or natural clearings in the thick almost tropical forests, others that he distributed the seeds among the settlers themselves to plant, and still others claim that in the damp land surrounding the marshes he established nurseries where he kept the seedlings until they were big enough to transplant. My Great-Aunt Mattie said that her father, who lived in rather a grand way for a frontier settler, had boxes of apples brought each year from Maryland until his own trees began to bear and then he always saved the seeds, drying them on the shelf above the kitchen fireplace, to be put later into a box and kept for Johnny Appleseed when he came on one of his overnight visits.

Johnny scattered fennel seed all through our Ohio country, for when the trees were first cleared and the land plowed up, the mosquitoes increased and malaria spread from family to family. Johnny regarded a tea brewed of fennel leaves as a specific against what the settlers called "fever and ague" and he seeded the plant along trails and fence rows over all Ohio. Some people said that he carried flower seeds with him to distribute among the lonely women who lived in cabins in clearings in the vast forest and that today the great red day lilies which grow along the roadsides or on the sites of old cabins, long disappeared, were spread by Johnny. They say also that Johnny sometimes carried in his "poke" as gifts tiny seedlings of Norway spruce which he gave to frontier wives to plant before their cabins. Both stories may be true for in our part of Ohio there is nearly always a pair of Norway spruce well over a hundred years old in the dooryard of every old house, and the red day lilies have gone wild in fields, on roadsides and along hedgerows.

In the next county there is an ancient apple tree which, it is claimed, was one of those planted by Johnny. I do not know whether this is true or not but I do know that in our pastures, on the edge of the woods and in the fence rows there are apple trees which are the descendants of those planted by Johnny. They bear a wide variety of apples from those which are small and bitter to those, on one or two trees, which are small but of a delicious wild flavor which no apples borne on respectable commercial apple trees ever attain. Their blossoms have a special perfume, very sweet and spicy, which you can smell a long way off, long before you come upon the trees themselves. They have been scattered here and there long ago by squirrels and rabbits and muskrat and raccoon who fed on the fruit of the trees planted more than a century and a half ago on the edge of clearings out of Johnny Appleseed's "poke."

And in our Valley, Johnny Appleseed is certainly not dead. He is there in the caves and the woodland, along the edge of the marshes and in hedgerows. When in early spring there drifts toward you the perfume of a wild apple tree, the spirit of Johnny rides the breeze. When in winter the snow beneath a wild apple tree is crisscrossed with the delicate prints of raccoon or muskrat or rabbit, you know that they have been there gathering apples from the trees that would never have existed but for crazy Johnny and his saucepan and "poke" of seeds. He is alive wherever the feathery fennel or the flowering day lilies cover a bank. He is there in the trees and the caves, the springs and the streams of our Ohio country, alive still in a legend which grows and grows.

Sometimes when I am alone in the old bottom pasture, or the woods, the memory of blind Great-Aunt Mattie returns to me. I can see her again, sitting by the edge of the clear flowing stream where the children left her, surrounded by cattle and the wild birds, her head a little tilted, listening. She has been dead for close to forty years and only lately have I begun to understand what it was she heard. It was the song of the earth and streams and forests of which Johnny Appleseed has become the patron saint in our country. It may be that while she sat there Johnny Appleseed was with her.

She had a verse which she used often to repeat to us children:

He prayeth well who loveth well
Both man and bird and beast

He prayeth best, who loveth best
All things both great and small
For the dear God who loveth us
He made and loveth all.

One hot summer afternoon when I was twelve years old we returned late to guide Aunt Mattie back to the farmhouse. As we approached the chosen spot the cattle were as usual lying in a circle about the place where we had left her. She was leaning against an ancient sycamore tree, her head thrown back a little, her eyes closed. My cousin and I thought she was asleep, but when we spoke to her she did not answer. It was my first sight of death and I felt no more terror than the cattle lying in the bluegrass in a protecting circle about her. It was all strangely a part of the Valley, of the whole cycle of existence and the most natural thing in the world. I remember that when I got back to the farmhouse I had an impulse to say, "Aunt Mattie has gone to join her friend Johnny Appleseed." But I was only a small boy and then it seemed silly. I only said, "Something has happened to Aunt Mattie." She was eighty-five years old when she died.

THE ANSON PLACE

But Johnny and Aunt Mattie have led me a long way from the Anson place, set there on the warm hillside. The house was just a square, comfortable house with no special architecture, yet with the kind of beauty which plain houses, built honestly with good country workmanship, always have. And it had an aura—the warm, hospitable aura which only houses have that are richly lived in. Clem Anson had been born in the house and brought his wife, who was a Shrack from the Valley, to live in it. Their five daughters had been born there and courted there and married there, and always since the beginning there had been young people coming and going. It was a gay, warm and friendly place, and below it on the hillside there was a rockery and garden filled with plants and shrubs and flowers collected by Mrs. Anson or given her by friends. It was a famous garden throughout that part of the state and in the long summer evenings people drove out from Mansfield and Mount Vernon and Ashland to see it from the roadside. It was remarkable not only for its beauty but for the mystery of how Mrs. Anson found time out of her busy life to tend and cherish it.

Clem Anson was a tall, lean man, a little stooped, with white hair and mustaches and the dignity and distinction of a man who loved his land and was proud of being a good farmer and a landholder. He had the kind of simple dignity which only those who love their land ever achieve. He might have died there instead of in the neighboring village where he retired when we bought the place, but for the fact that in the same year the youngest and last of the five daughters married and left home. He was old and tired and with his daughters gone, the place was lonely. He was a part of my early memories of the Valley for when I saw him again after thirty years, I remembered him as the farmer who never resented the presence along the bank of the creek or in his fields of a troop of small boys playing Indians or going on a snake hunt. Even in his seventies his eyes were blue and bright with the youth of very old people who like very young people about them.

The Anson place had a character and a feel about it quite as strong as the Ferguson place but quite different. There was nothing remote or solitary and magnificent about it and certainly nothing *fey* or haunted. It was human and alive with people coming and going all the year round up and down the short, steep lane. It was the Anson place we chose as home. It was a good choice. The spot was already warm like the nest of a rabbit, and since we came there the tradition has gone on unbroken. People still come and go all the year round, up and down the short, steep lane. There are three daughters instead of five, but the house is always filled with young people and, no matter how many there are in the house, there is still an extra bed, and an extra place at the table.

Halfway down the Hastings Road below the Anson place near the little river that winds through the Valley there stands a great water elm, the biggest perhaps in all the state of Ohio. Out of its very roots flows a spring and its huge trunk measures twenty-six feet around the base. Its vast drooping branches completely cover the miller's house, hanging nearly to the ground.

The miller's house is small and sits a little below the road beside a big spring. It should be a haunted house for four people were murdered in it, but it is not. It is a singularly cheerful place perhaps because it has a lovely view up the hillside toward the Ferguson place and down the Valley across the deep rich green bottom pastures where the dairy cattle and the horses feed. And when you are in the miller's house the sound of running water is always in your ears.

The old mill once stood on the opposite side of the creek but it has long since disappeared. Its great timbers form the framework of the big feeding barn where the beef cows and their calves are sheltered from winter storms; the big hand-cut stones of its foundation form the steps and the chimneys of the big house that grew over and around the old square house of the Anson place.

It was a girl called "Ceely" Rose who lived in the miller's house and poisoned there her father, mother and two brothers. Her father was the miller and the tragedy occurred before I was born but it is a legend of the Valley and as a small boy I sometimes heard the grownups talk about "Ceely" Rose. Her father's name was Alexander Rose, a good name for an honest miller, and "Ceely" had been christened Cecilia. The family didn't belong in the Valley; they had come there from the hills of Tennessee.

In all the years I was away, the vague story faded in my memory until it was completely forgotten. It returned one morning a little after we came to live on the farm when an old man called Mr. Charles drove over from Bellville to inquire after my father. And while we stood talking on the bridge a little way from the miller's house, Mr. Charles looked up suddenly and said, "Oh, why that's 'Ceely' Rose's house."

And the name I had not heard for perhaps forty years became alive again dimly in the remote recesses of my childhood memory. I said, "'Ceely' Rose! I remember the name but I don't remember who she was.

And then old Mr. Charles told me the story.

From the very beginning Ceely was an unfortunate, tragi-comic figure, a big, strapping girl, round-faced and heavy, not only in body but in wits. She was not very bright in school and sometimes in the red brick schoolhouse at the end of the Hastings Road, where she lingered behind in class with children smaller than herself, the other children teased and made fun of her. And each day as she walked home along the Hastings Road and across the bridge past the Fleming place she used to see the Fleming boys, sometimes in the front garden, sometimes in the fields, sometimes doing the chores. One of them called Hugh was a handsome boy, a little older than Ceely, and sometimes Ceely would stop and, leaning on the fence, call out to him. It embarrassed Hugh for the other children would stop and mock him and poor Ceely. The sad thing was that Hugh was not only handsome but he was kind, and instead of sending Ceely on her way with some loutish joke, he would stop plowing and talk to her. Because there was one person who was kind to her, Ceely, big and overdeveloped and dim in her wits, fell in love with him.

Her attentions embarrassed the boy, especially when it came to the point where she not only stopped to lean on the fence and talk to him but, whenever she was able to escape her mother's watchful eye, she ran away to find him in the fields or in the barn. The boy, aware that the young people laughed at him over Ceely's passion, tried to be rid of her, but still he was kind. He told her that she was too young and that if she came to see him, it would get him into trouble with her parents. At last he succeeded in evading her, but the poor dim-witted girl brooded over her thwarted love and slowly out of his casual remark and the restraint exercised by her parents, grew an obsession—that all that stood between her and young Hugh was her parents and brothers. Slowly she came to believe that if she could rid herself of them, she could have her Hugh.

And so with the directness of a mind dimmed and twisted, she soaked the arsenic out of flypaper and one morning went down to the springhouse and mixed it into the cottage cheese which her father and mother and brothers ate for breakfast. Before noon her father and one brother were dead and her mother and other brother desperately ill. They buried the father and one brother but no one could discover how the family had been poisoned and how Ceely herself escaped. Suspicion pointed to her but when they questioned her they could discover nothing and on the face of it there seemed to be no motive.

But Ceely in her confused mind was both shrewd and determined. For a long time her mother and brother were ill and then one morning, just as they had recovered, Ceely repeated her performance, doubling the dose of arsenic to make certain, and this time the mother and remaining brother died and Ceely was left alone. But Hugh Fleming, alarmed by all that had happened, had gone away.

By now, suspicion of Ceely's guilt had grown in the minds of the whole county, but there was no way of proving that some neighbor with a grudge had not slipped into the springhouse during the night and poisoned the cottage cheese which had wiped out Ceely's parents and brothers. That was Ceely's own theory; she stuck to it, even going so far as to name respectable farmers in the Valley who at one time or another had quarreled with the miller. She was shrewd and her dull, devious mind was not to be shaken by any amount of questioning.

It was the sheriff himself who finally got her confession. They allowed her to live on alone in the house where all her family had been murdered, a position which did not appear to trouble her. It must have been an odd life, alone, regarded with suspicion by all the Valley. Sometimes, driven by loneliness, she paid visits to the neighbors but, remembering the arsenic, they did not ask her into the house but only humored her and sent her away again. She told them all that Hugh Fleming was coming back to marry her. She said that was why she stayed on in the lonely, haunted house, so he would be certain to find her when he came back.

And then one day a girl called Vilma Smith who lived in the Village at the end of the Valley and also came from Tennessee asked Ceely to come and spend the day with her. Poor Ceely snatched eagerly at the kindness and early in the morning Vilma herself drove up the Valley to fetch her. It was a summer day with the heat hanging over the Valley and after a big midday dinner, Vilma suggested to Ceely that they go out to the cool barn and have a nap in the freshly cut hay.

While the two girls lay there talking drowsily, Vilma began to confide in Ceely the story of her own thwarted love. She said that she loved the son of a neighboring farmer and that he loved her but that there was a feud between the two families and that her parents refused to let her marry him. Lying there in the sweet-smelling hay she told Ceely that she had decided there was only one course to take and that was to poison her family and free herself of them forever.

When she had finished the tale, she asked Ceely what she would do in the same circumstances, and after a moment Ceely said, "I'd kill 'em. That's what I'd do. That's just what I did." And slowly, bit by bit, she told the whole story of soaking out the arsenic and putting it in the cottage cheese.

When she had finished there was a little silence and then from the cow barn, just below the mow, the sheriff and the prosecuting attorney appeared to say, "Cecilia Rose, in the name of the law I arrest you."

There was a trial and there was some dispute over the legality of the fashion in which the confession had been obtained, but in the end there was no doubt as to Ceely's guilt. She was sent away to a hospital for the criminally insane. The mill wheels of the big mill turned no more and the miller's house stood empty until Clem Anson bought the whole place, used the miller's house as a tenant house and pulled down the old mill to build the great cattle barn.

The year after we came to the Valley, poor Ceely died in the asylum. She was eighty-three years old.

So that was Ceely's story as old Mr. Charles told it. That is what lies behind the record of the deeds transferring the site of the mill and the miller's house from John Ferguson to Alexander Rose, and from the heirs of Alexander Rose to Clem Anson.

And while old Mr. Charles and I stood there by the bridge he told me another story of the Valley—the story of the highwaymen. It came about naturally for it concerned the Ferguson place, remote and lonely against the sky, and the two old bachelors and the old maid who were the last of the Fergusons.

Before the days of automobiles and telephones the Valley and all the county below it was remote and lonely country, not very different from the English countryside where highwaymen flourished at the end of the eighteenth century. The roads wound through thick woods and in those days there were still many covered bridges ideal for holding up passers-by in buggies or on horseback.

At the beginning of this century there appeared in our county and the neighboring county a band of highwaymen, three in number, who operated in the traditional manner, sometimes on horseback, sometimes on foot. They would appear suddenly on a covered bridge or out of the woods on a lonely road to hold up a farmer on his way home from market. There were two tall men and a short one and all three wore handkerchiefs over their faces and hats pulled down over their eyes. They carried an old-fashioned bull's-eye lantern and their victims, blinded by the light, never saw their faces or had a good look at the horses they rode. For months they terrified Pleasant Valley and the neighboring valleys until no farmer would venture to go anywhere after dark.

When there were no more victims on the road, the bandits adopted new tactics, descending after midnight upon lonely farmhouses to hold up the farmers and their wives and take whatever money there was in the house. If the victim resisted or declared he had no money, they sometimes tortured him by holding lighted matches or hot coals to the soles of his feet until he admitted where the money was hidden. That was what they did to Euphemy Ferguson. Long after midnight, they made their way on foot up the deep ravine with the cave and waterfall to the little house where the two bachelors and the old maid lived beside the spring, and while the two brothers, tied and gagged, looked on, they burned the soles of Euphemy's feet until she told them where the meager eight dollars and some coins were hidden. The next day they found the footprints of the highwaymen in the deep snow at the bottom of the ravine.

And meanwhile, the sheriffs of the two counties organized posses which patrolled the roads at night, but they never found the bandits or any trace of them. The odd thing was that on the nights they patrolled the roads the highwaymen never appeared, and so the legend grew that someone in the posse was in league with them or that the highwaymen on their off nights rode with the posse.

Neighbor began to suspect neighbor and vicious whispers went the rounds of the two counties.

And then one day there appeared at the office of the sheriff of Richland County a woman who said her name was Mrs. Hanby and that she lived in the wild country south of the Valley. She had a story to tell. Her husband, she said, was one of the highwaymen but he had left her and was living with another woman. He was the small bandit. She had had enough of the whole thing and had come to confess. The other two bandits were the sheriff of Ashland County, who led the posse, and his brother. They were the tall ones.

They went to jail for twenty years and after that the Valley settled down again into peace and security, but for the rest of her life Euphemy Ferguson was a little queer and walked rather like a duck.

THE FLEMING PLACE

The Fleming place never had much character and you never felt about it as you felt about the Ferguson place and the Anson place. Even today it remains rather nondescript and uninteresting, just a farm, any farm. It had no Norway spruces in the dooryard, no ancient lilacs, no spring near the house but only a drilled well. There was never any feeling that the place had ever been loved and cherished like the Anson place or that it belonged to another world like the Ferguson place.

Potentially, the Fleming place was the best farm of the three. The gently rolling bottom fields were of alluvial or glacial origin, the kind of gravel, clay, loam which a wise farmer would cherish and delight in working. It needed no artificial drainage. There were springs on it and even an artesian well in the bottom pasture that poured out a three-inch stream of icy water. But no one had ever farmed it well, even in the old days. No one had ever cherished or fed that potentially rich soil. The tradition which had guided the farming of the Fleming place for more than a century was the evil tradition which has dominated most farming in America. The men who had owned or rented the Fleming place had not farmed it; they had "mined" it, as if its rich fields were no more than coal or iron ore to be dug up and sold.

And so, from the time of the first settler, its fertility had slipped down and down under the greed and ignorance of the men who worked its gently rolling fields. Clem Anson had loved the earth beneath his feet and the Fergusons had loved the wild beauty of their farm against the sky, but nobody had ever loved the Fleming place and it had no soul save for the acre lying on a knoll above the little creek where all the early settlers of our part of the Valley lay buried. No burial ground can be soulless and the white fence which surrounded it had protected the soil where the old people lay buried against the miserable greed which had very nearly destroyed the rest of the farm. Within the little enclosure the trees, the shrubs, the periwinkle grew lush and green.

Once there had been on the Fleming place a lovely old house, long and low with a gallery all along the southern side. It was one of those happy houses which appeared to have grown out of the earth itself. But no one had cherished the house. In all the years of its existence it had never been painted and when I last saw it as a boy, it had weathered the silvery color of an ancient fence post. In spring and early summer it was half hidden among clumps of lilac and mock orange. That house had had a soul but with neglect it had rotted slowly until the owner at last found it was no longer livable. He pulled it down, and in its place he built the ugliest house in all the Valley.

I do not know where the money came from for the new house. I doubt that it came from the farm itself, although much of the timber came out of the woods on the hillside opposite and some of the beams came out of the old house which had been allowed to die through neglect. The ancient lilacs and mock orange were rooted out to make way for a monstrosity built upon a plan furnished by a mail-order house. It was neither a simple house built carefully and honestly by country carpenters, nor a beautiful house like many in the Valley modeled upon the fine old brick houses of Pennsylvania and Maryland. It was simply a big, pretentious, vulgar house built for show, without taste or knowledge or any sense of beauty whatever, a monstrosity among the hills and meadows of the Valley. And certainly no one ever loved it. Never was there a more soulless house. The inside was practical and compact—the mail-order house architect was good enough on that side—but one could live in it only because while *inside* the house one could not see the outside. What one saw out of the ill-proportioned windows was a long view up and down the Valley, a view so beautiful that it made one forget the house itself.

It had a long series of tenants, most of them not farmers at all, and at last people who worked in the town, rented the house not because they loved the land but because the rents were cheaper in the country. And the fields had reached the final infamy, that last step before a farm dies; they were rented out to neighboring farmers who, having no interest in land which they did not own, took everything from it and put nothing back.

Although we lived happily enough in that house for eighteen months we never became attached to it. Some day, I hope, when taxes permit us to live again, we shall be able to tear down the house or move it away and build in its place a house which will not be an eyesore to the whole of the Valley, a house which will be a farmer's house and not a pretentious suburban monstrosity.

That, then, was the history of the three farms that came to be joined together again as one farm after a hundred years. A lot of people thought we were foolish to waste good money on farms in hill country, two of them exhausted through years of bad farming. They said that if they had had that much money to spend they would have spent it on the rich, deep, flat land of northwestern Ohio.

There were at least three reasons why we bought these farms. One was that I loved Pleasant Valley and had never been able to escape it. Another was that life in a flat country was intolerable to me. As a Valley neighbor once put it scornfully, "In flat country, there is never any place to go. There's no reason ever to leave home because wherever you go it is exactly like where you came from. In hill country there's a new world over the crest of every hill and you never know what's goin' on over there until you pass the crest of the hill and come down in the valley on the other side."

But there was a third reason, more profound than either of the others. There were things I wanted to prove; that worn-out farms could be restored again and that if you only farmed hill country in the proper way, you could grow as much as on any of the flat land where something rich was lacking from life. An old friend of mine said I was buying "some fine scenery." I didn't argue that; there was no doubt of it. You could never find the lonely beauty of the Ferguson place in flat country. And I knew that wherever I had been in the world, hill people were quite different from people who lived in flat country. They were freer, and wilder and more colorful.

And there was a kind of challenge which I found irresistible, a little I think, like the challenge with which the wilderness confronted the first pioneers. In the space of a little over a century those first pioneers and their descendants had passed over the surface of America like a plague of locusts, "mining" and destroying the land as they went, until at last they reached the Pacific Ocean. And then suddenly there was no more free and virgin land to destroy. That's all there was; there wasn't any more. And slowly, imperceptibly, the fact of that disaster began to make itself felt in the economy of a whole great nation. The shortage began to make itself felt in a living standard slipping slowly downward to the level Europeans had known for a thousand years. People didn't know about what was going on. Neither farmers or city people. I knew perhaps better than most because I had seen over the whole of the world what had happened to nations when their agriculture grew sick and their soil impoverished. What happened was first economic sickness and finally death, not only of agriculture but eventually of the nation and its civilization.

I knew in my heart that we as a nation were already much farther along the path to destruction than most people knew. What we needed was a new kind of pioneer, not the sort which cut down the forests and burned off the prairies and raped the land, but pioneers who created new forests and healed and restored the richness of the country God had given us, that richness which, from the moment the first settler landed on the Atlantic coast we had done our best to destroy. I had a foolish idea that I wanted to be one of that new race of pioneers.

I might wreck myself. I might break my heart, but so had many of those first pioneers. Anyway I didn't want a dull life in a flat country, nor, fortunately, did any of my family.

THE PLAN

IT WASN'T A JOB I COULD DO ALONE. I HAD BEEN AWAY FROM America too long, far too long out of touch with the machinery of agricultural administration, of American agricultural experiment and progress. In some ways I was far more familiar with what was going on in France, in Sweden, in Syria or even India. I was born in a small town where one only had to walk a few minutes in any direction to be in the open country. I had spent half my childhood and boyhood on my grandfather's farm—*The Farm* of which I had once written all the history in the most intimate possible detail. At sixteen I had gone off to Cornell Agricultural College meaning to be a farmer. I might have gone on with that career but for two circumstances, both extremely powerful. One was a potent urge to become a writer; the other was my mother who was determined that I should not, as she phrased it, "waste my life on a farm."

She was herself the daughter of a skilled farmer, the first Master of the Grange in the state, but in her heart she always carried a bitter resentment of the fact that a piece of land anchored one forever to one spot. She had in herself the seeds of the restlessness which made it necessary for me to see the whole world before I settled down. She always counseled me to be a writer first and then at my leisure become a "farmer." I knew that "a gentleman farmer" was what she meant even then—a man who had a big income from investments, established a "model dairy" or had saddle horses in mahogany stalls. I never had that in mind, even then. It is still less in mind today.

During the first year at agricultural college my grandfather died at a great age and my mother insisted that I come home and operate his farm myself for a year in order to make a decision. Because she was a "powerful influence" I yielded and ran my grandfather's farm, not too badly, for a whole year and learned what she meant by being "anchored to a piece of land." There were twelve cows to be milked twice a day, hogs, horses, chickens to be fed, hay to be made, not when one chose but when the hay was ready and the weather right. At the end of the year I knew that it was no good, that I could never settle down until the restlessness was satisfied. At the end of that year, I did not return to the college of agriculture; I went instead to the Columbia School of Journalism. But even that did not last for long. Before the first year was finished I went off to the war. I never gave either university much of a chance to teach me anything, but I never learned anything at all at either one.

I became a writer and I made lots of money and my books were translated into every language in Europe and even into Chinese and Bengali. I don't think my mother could have asked for more, but for all that, the old itch for the land never died. In all my life I lived only three years in a city and they were the three most unhappy years of my life, two of them in New York and one in Paris. They were unhappy because I was bored, despite all the distinguished and celebrated people I knew, and all the supposedly exciting events in which I participated. Always there were two or three hours a day when I did not know what to do with myself. If I had had a weak head or any taste for drink I could have slipped down the path so many writers have gone out of boredom. I just wasn't born for city life. On a farm, if you are born for it, no day is ever long enough to accomplish all there is to be done. No city can offer any excitement comparable to what happens when there is a new pure-bred calf or the whole landscape comes alive with the change of season. No excitement can equal the slow satisfaction of witnessing a tired, worn field come back to life and fertility.

And so during all the years away from my country I came to know farmers and agricultural stations in every part of the world, and I learned much, many things that we have not yet learned in agriculture even here in America. And always I had, save for three years, a piece of earth in which I could work, watching and learning things about earth, water, air, plants and animals.

. . .

The first step was obviously to find myself a partner in the undertaking, a younger man who knew all the developments of recent years, who was a farmer and was in touch with the agencies of government, of the extension school, of the agricultural colleges which made easily accessible to the farmer all the knowledge concerning agriculture which they had been piling up for a long time. There were a hundred applicants for the job of taking over and restoring John Ferguson's section of hill and valley land. They were of all kinds, young and old and middle-aged, some of the products of agricultural colleges, some rabid theorists, some just plain old-fashioned dirt farmers. They came from all over the country. The selection was a long process and I learned much from talking to all those men. But none of them seemed right; either they knew too little of recent developments in agriculture or they knew too much and had too little contact with the earth itself. Or they were too old and "sot" in their ways, or they were too young and too eager, too innocent of the hard work that is inevitably a part of farming.

A good farmer in our times has to know more about more things than a man in any other profession. He has to be a biologist, a veterinarian, a mechanic, a botanist, a horticulturist, and many other things, and he has to have an open mind, eager and ready to absorb new knowledge and new ideas and new ideals.

A good farmer is always one of the most intelligent and best educated men in our society. We have been inclined in our wild industrial development, to forget that agriculture is the base of our whole economy and that in the economic structure of the nation it is always the cornerstone. It has always been so throughout history and it will continue to be so until there are no more men on this earth. We are apt to forget that the man who owns land and cherishes it and works it well is the source of our stability as a nation, not only in the economic but the social sense as well. Few great leaders ever came out of city slums or even suburbs. In France, in England, in America, wherever you choose to turn, most of the men who have molded the destinies of the nation have come off the land or from small towns. The great majority of leaders, even in the world of industry and finance, have come from there. As a nation we do not value our farmers enough; indeed I believe that good farmers do not value themselves highly enough. I have known all kinds of people, many of them celebrated in many countries, but for companionship, good conversation, intelligence and the power of stimulating one's mind there are none I would place above the good farmer.

But there are two other qualities, beyond the realm of the inquiring mind or the weight of education, without which no man could be a good farmer. These, I believe, are born in him. They are a passionate feeling for the soil he owns and an understanding and sympathy for his animals. I do not believe that these traits can be acquired; they are almost mystical qualities, belonging really only to people who are a little "teched" and very close to Nature itself.

Often enough people discover late in life that they have these qualities, without ever having known it. They did not acquire them suddenly; they were always there. It is only that through the accident of a fishing trip or the purchase of a farm, they discovered them. I have any number of friends who spent all their lives as bankers and industrialists or workingmen or insurance salesmen, only to discover at middle age that in reality they were farmers all the time, without knowing it. I know of no human experience more remarkable than that of men whose whole existences are changed and enriched by the discovery late in life that they have a close bond with the earth and all living things, and that they have lost vast and intangible riches by not making the discovery when they were younger.

Conversely, there are many men on farms in America who have neither that love of soil nor of animals. They are the bad farmers who have done us such great damage as a nation. They do not belong on farms. They are there, most of them, because they were born there and have not the energy to quit and go to the cities and factories where they properly belong. There are too many of them in America, and they have cost us dear.

For the good farmer, his animals are not simply commodities without personality destined only to be made into pork chops or beef steaks or to produce milk all their lives. To a good farmer, each animal has its own personality. A good farmer cannot himself sleep if his animals are not well fed and watered and bedded down on a cold winter night. Watch any good farmer showing his sheep or cattle or hogs at a county fair or an international stock show and you will understand how much he respects the animals that are linked into that chain of life which explains and justifies the whole of his activity. Or watch any 4-H club boy or girl with tears in his eyes when the moment comes for him to part with the fat steer he has raised and brought to a cattle show. He has slept in the straw in the stall beside his steer for days. The steer is a part of the richness of his own existence. He will go cold himself or go without food and water before the steer shall be deprived of these things.

Looking for all these qualities among a hundred or more strangers was not an easy job, but I found them at last in a young fellow called Max Drake. He was thirty-two years old, with a wife and son. He was himself the son of a good farmer. He had a brilliant record in the state agricultural college. He had worked with 4-H clubs and been a substitute county agent. He was interested in anchoring the soil and salvaging what remained of our good land. And he was interested also in the whole intangible side of a farmer's life which had to do with farm institutes and square-dancing and fun.

We made a deal and Max and his wife and son came to live in the miller's house where "Ceely" Rose had done away with her parents and two brothers. Fortunately they were not superstitious people. They liked the little house with its big spring and its huge elm tree and the wide view up and down the Valley across the bottom pasture where the dairy cows fed. It was still winter when they came there.

As winter broke into spring we sat night after night beside the big stove in the miller's house or in the big ugly house on the Fleming place where I came with my family to live until the changes had been made in the old house above us on the side of the hill. I think that for both of us, there had never been or ever will be a more stimulating experience than the working out of the Plan. We had between us six hundred and forty acres of woods and pasture and farmland and springs and streams, a small kingdom which we sought between us to bring back to life. It was a little like planning the recreation of a world of our own, secure and complete and apart.

The Plan grew slowly at first and then more and more quickly out of the lives, the education, the experience of both of us. Both of us knew pretty definitely what we wanted to do. There were disagreements and compromises but in the end the Plan emerged. It was something like this:

There was first of all the soil itself which was the foundation of our own well-being and security as it was of that of the whole nation. Much of it was already gone, washed off our hills, down Switzer's Run, into the Clear Fork and thence into the Muskingum, the Ohio and the Mississippi, where so many billions of tons of our good American soil have gone since first we cut off the forests and plowed up the grassbound prairies. We had first of all to stop that destruction and heal the gullies which made ugly scars in every sloping field. And we had to bring back to the soil the fertility which, save on the Anson place, had been dissipated by ignorant and greedy farming.

We had to remodel the big old barns built in a more leisurely time when labor was cheap and the rich soil had not yet been raped of its fertility. Those barns were planned too before the days of machinery and to operate them it took too many man-hours, more hours than any farm could afford and succeed in a time when labor costs had doubled and tripled.

And there was the question of livestock which was vital in the production of manure to bring back the fertility of the wasted fields. We decided that whatever we sold off the farm would have to be able to walk off, a sound principle of any farm, rich or wasted. No straw, no hay, and a minimum of grains would be sold off the place. It would be feed and the animals who ate it marketed, leaving behind them the tons of manure which spells wealth to a farmer anywhere in the world.

And there was to be a whole program of green manures, the clovers and other legumes, rye and rye grass to be plowed under to restore both nitrogen and humus to the tired,worn-out soil. And there was the question of lime and phosphorus and potassium and other minerals and trace minerals which had been leached out by erosion and removed by greedy "mining" of the land—all these minerals without which plant life remained sickly and feeble and animals which fed upon it small-boned and anemic.

We knew, too, that poor worn-out land made not only poor crops and scrubby cattle; it made poor, underdeveloped, undernourished people as well. That was one of the great problems of American agriculture—the decline of human stock through the decline in fertility and mineral content of the soil itself. *"Poor land makes poor people"* is a saying every American should have printed and hung over his bed.

And there was the question of the timbered land which had gone the way of most farm wood lots in America. Long ago most of the bigger trees, save only the sugar maples, had been stripped from most of the land for quick cash profits. Into what was left of the woods sheep and cattle had been turned to feed for many years. The pasture was poor and scrubby and the cattle and sheep got little or no nourishment from it, and in their hunger they had eaten off the young seedlings year after year until there was no new crop of trees coming on to supply the farmer and the nation with timber. The men who had farmed these three farms were getting neither pasture nor timber from the wood lots and as most of the woods were on steep ground unfit for farming, the soil was washed badly in spots and gullies occurred even on the wooded land. To check all that we planned to shut out all cattle from the wood lots and give the forest a chance to seed itself and become productive once more.

There would be other benefits as well. A forest floor green once more with seedlings does not permit the water to rush across it, cutting gullies, carrying away topsoil and encouraging the rainfall to escape into the Mississippi and the Gulf of Mexico instead of staying in Ohio where we needed it desperately for crops and springs and industrial and urban use.

There was too the question of restoring the native pastures of bluegrass and white clover, perhaps the finest pasture for livestock in the world. Our bottom pastures which lay along the little creek were alluvial and deep-soiled and most of the land had never been cultivated since the trees had first been cut away; but in the course of years both lime and phosphorus as well as other minerals and trace minerals had been leached away even there or consumed and carried off in the bodies of the animals which fed off it. On the hill pastures the topsoil was often gone along with the minerals and where there should have been bluegrass and white clover, there was only poverty grass and the tough wire grass that moves in on poor land. In many of the pastures the animal droppings of years provided nitrogen, but nitrogen alone is not enough to produce good grass. All the pasture land would need both lime and phosphorus, not only in order to grow better grass, but to feed these minerals into the bones and sinew of cattle, horses and sheep, and the growing children on the farm.

As the new pioneers bent upon restoring the land, we had to put back by every manner of ingenious means the very elements the first pioneers had removed recklessly or permitted to disappear through erosion and neglect. The problem was, on the surface, a simple one but both Max and I knew that it was not so simple as it appeared to be. Working *with* Nature, we would be recompensed by her sympathy, but beyond a certain point Nature could not be pushed. To build the precious topsoil which had been treated so carelessly by our predecessors, it had needed millions of years in the natural process of the disintegration of rock and clay and rotting vegetation. With the aid of the knowledge and experience of agricultural civilization, we knew that we could restore the soil infinitely more quickly but only if we worked *with* Nature rather than *against* her as our predecessors had done.

One thing was in our favor. All our land was glacial moraine, of the soil type known as "Wooster"—great hills of gravel, sand and clay, scraped from the surface of the land north of us for a distance of fifteen hundred miles north to the borders of Hudson's Bay. Some of it, alluvial in character, had been washed from beneath the vast glacier as it began to disintegrate with a softening of climate that had happened a million years before John Ferguson came into the Valley. Mixed in these hills, all the way down, were gravel, lime, phosphorus, potassium, fluorine, iodine and countless other elements. It was good stuff to work with, to build upon. In creating new topsoil, the process required most of all masses of decaying organic matter to hold the moisture, the volatile nitrogen and anchor the minerals we planned to use in the replenishing process.

In our case when the topsoil had disappeared we did not come upon rock, or hardpan, or blue clay as happened in the case of so many thousand worn-out farms in America. When that happened, a farm was really dead and gone, a total agricultural loss, fit only for reforestation or to be abandoned as man made desert. Both Max and I knew that millions of acres of good American land, based upon hardpan or shale, had been reduced to desert conditions. It could never be restored quickly by any process now known to man. We were lucky—where the dark topsoil was lost we came down to glacial drift, made up of many kinds of soil stirred together by the great glacier. We had something to build upon.

But the plans we had in mind were not concerned alone with the soil. They had social and economic aspects as well, for we both knew that agriculture in America was sick, with a wasting illness which no amount of subsidies or superficial measures imposed by a highly centralized government could cure. We knew too that when agriculture is sick, the illness in time pervades the whole of the economic structure of the nation. That much was history, unarguable, confirmed by the story of more than one dead or decaying nation since the beginning of time.

Some of the reasons, like the wasting of the soil, lay with the farmer himself, born of ignorance or greed or carelessness. We knew the evil of crop systems prevailing in the areas where only one crop—tobacco or wheat or cotton or corn—was raised year after year for decades while the soil grew thinner and thinner and more and more exhausted. That problem did not concern us for Ohio country was given over naturally to general farming and livestock. But out of that single-crop system, out of the laxness of thousands of farmers and their wives, out of the mechanization of our civilization, had grown another abuse which played a large part in the sickness of agriculture. The old economic independence of the farmer, his sense of security, that stability which a healthy agriculture gives to the economy of any nation, had broken down, and one of the most important reasons for the breakdown was the farmer's dependence upon things which he purchased rather than producing these same things off his own land.

I had memories of the farm of my grandfather, a farm which in itself was a fortress of security. On a hundred acres he had raised and educated eight children, strong, healthy, constructive citizens. He had bought little more than salt, coffee, tea and spices. All else he had produced out of the earth he owned and cherished, and when winter came there was a great cellar stored with home canned goods, and a fruit cellar heaped with apples, potatoes and all kinds of root vegetables. In the dry-cold attic hung rows of hams and flitches of bacon. In that house I ate as well as I have ever eaten anywhere in the world. I knew that he had spent little or nothing upon food for himself, his wife, his eight children, his hired man and girl and all the relations who were always staying in the big house. At the end of the year all these vast quantities of food showed up as profit on the right side of the ledger.

Opposed to that memory was the knowledge that scattered over the whole of our nation were thousands of farms where the farmer and his wife bought the bulk of their food out of cans off shelves in town grocery stores, very often in the very midst of the gardening season. The extreme examples were those one-crop area farms where wheat, or corn or tobacco or cotton grew up to the front door and the farmer's wife bought canned peas and beans in June or July or August. To our astonishment, we found during the first year of operation, that neighbors from farms, good farms, with two hundred or more acres of land, came to us to buy apples because they had not even one apple tree on their land. This happened even in our rich diversified Ohio country.

The economy of a piece of land is at once a complicated and a simple economy. The farmer's land is his capital. If he does not produce all that he can produce within economic and climatic reason, he is failing to utilize his capital to its full advantage. Whatever he buys that he is able to produce can only appear on the red side of the ledger at the end of the year.

On the point of self-sufficiency, Max's faith was less strong than my own, I think because his faith in machinery and in such economic pitfalls as the installment plan was greater than my own, and because he had never seen in this rich country the things I had witnessed and lived through in countries where there were inflation and food shortages and rationing, and discord and civil war. I was perpetually haunted by the terrible economic insecurity which a mechanized and industrial civilization imposes upon the individual.

I told Max about these things and he listened with intelligence and interest but even though he had witnessed the misery of a great depression, he never quite believed that they could ever come to rich America. Even while he talked there by the big stove in the miller's house, rationing, food shortages and other hardships were already in the cards for America. Even disorder and civil conflict were not beyond the realm of the imagination after the war. If and when such things did come, I wanted, I told Max, to be on my own land, on an island of security which could be a refuge not only for myself and my family but my friends as well. I did not want to be in Detroit or Pittsburgh or any great industrial city. I had lived through inflation, strike, violence, shortage and civil disorders three or four times in Europe and I did not believe that we, as a nation, were any more immune to the results of our folly than any other nation. But Max, like so many other Americans, did not believe that we could ever be involved in the disasters which afflicted the rest of the world.

Because he was lukewarm about the self-sufficiency angle, that part of the program fell wholly to me. I knew what I meant to do. I meant to have as nearly everything as possible, not merely chickens and eggs and butter and milk and vegetables and fruit and the things which many foolish farmers buy today. I meant to have guinea fowl and ducks and geese and turkeys which could live off the abundance of the farm without care in a half-wild state as such birds live in the *basse-coeur* of a prosperous French farm. I meant to have grapes in abundance and plums and peaches, currants, gooseberries, asparagus—indeed a whole range of things which could be had for the mere planting and the expenditure of a little work and care.

One pond already existed on the farm and with the rolling land and the abundance of springs and small streams the building of ponds became a simple enough procedure. I meant to include ponds which would produce a constant supply of fish. And I meant not only to operate the maple sugar bush again but also to have bees which not only would fertilize crops but produce hundreds of pounds of honey. If there was too much fruit it could be given away, or sold at a roadside market or left to the birds and the raccoons.

I have never believed in the superficial folly of Henry Wallace's program of scarcity. There has, in fact, never been a surplus of food for the people of the world. There has always been a scarcity, with surpluses piling up in some countries and at times even in states and cities, because distribution has been abominable and because ill-managed trade relations between nations have crippled the whole system of distribution. The fundamental solution of the more grave economic problems of agriculture lies not in burning coffee and killing little pigs but in getting agricultural products to the areas where there is not only a market for them but where often enough they are desperately needed.

And thus self-sufficiency, another expression of the ultimate security, became an important and integral part of the Plan which was eventually to bear rich dividends not only for myself and my family but for all the other families on the farm.

There was another reason for the sickness of American agriculture which troubled both of us. This was the gradual disappearance over large areas of the family-sized farm and establishment of great mechanized farms which were more like industries than farms. The pattern repeated itself in many forms, superficially different, but essentially the same, supplanting a sturdy family by semi-industrial, often imported labor. Neither Max nor myself believed that the impulse of our times toward regimentation, centralization, mechanization and industrialism necessarily represented progress. It seemed to us and still seems to me a dangerous path with "rewards" at the end which are bought at the cost of physical stamina, human decency and the dignity and self-respect of man. We knew that the greater part of our migratory population in America—one of our most serious economic and social problems—came either from worn-out farms or from land that had been converted into factories in the fields. This roving population was one that was certain to increase in proportion to the increasing destruction of soil and the industrialization of the remaining farm lands.

We knew too that in these times the high cost of farm machinery and the necessity for it made it virtually impossible any longer for a young man to marry, purchase a farm and equipment and establish a family. If he inherited farm or machinery, he might succeed in getting along even with all the handicaps imposed by a sick agricultural economy, but if he had to buy both farm and machinery, the situation was impossible. Without efficient modern farm machinery he could not hope to succeed.

We sought a way to operate a big farm without dispossessing families. We sought a way of helping a young fellow to get started in life. We sought a way of raising the standard of living of all of us on that farm. Both of us, Max, out of his experience in this country and I, out of my experience both here and in a half dozen countries abroad, had faith in co-operative effort as the solution to many of the illnesses of agriculture, and together after many nights by the stove in the miller's house we worked out that part of the Plan.

We took as our model the collective farm as it had worked out in Russia. We accepted its principles only in a large and rather loose way, in which I, myself, as the capitalist, was substituted for the state. Both of us had faith in the belief that some sort of a cooperative farm could be worked out under proper management so that the income and living standards of all the families involved would be as good or better than that of the average-sized family farm and far better than those of the average moderate-income city dweller.

Under the Plan each family would have a house, rent free, with light and heat, bathrooms and plumbing, all its living save only coffee, spices and sugar, as well as a salary above the average. In my role as capitalist or "the state" I agreed to put up the money on the adventure exactly as I would invest money in a factory or a business project. I would assure the finances until we came to the point where we turned a profit. From then on I should take the first five per cent as a sound but not exaggerated profit on the investment and as an offset against the salaries and living expenses of the others. Once this five per cent was paid off annually, any profits above that amount were to be divided pro rata according to the salaries paid each worker, which varied according to their education, skill and value to the common enterprise.

I was to share as well in this distribution of the profits above five per cent, my share being based upon the pro rata share of the manager of the enterprise. It was recognized that while I took the risks, I also benefitted by the enhanced value of the property as we built up the fertility of the soil. The growing fertility, however, benefitted by increased production, and profits the others participating in the co-operative. It was also recognized that I contributed knowledge not only of agriculture but also of world and market conditions with which my profession and political and economic interests constantly brought me in touch.

And finally there was the question of the house that was to shelter my family, I hoped, for the rest of my life. During twenty years of married life we had never really had a house of our own. We had lived for a time in houses on the Basque coast, in Switzerland, in Kensington, in Long Island, in Baroda, in Mysore. We had a liking for houses and always in a place which pleased us, like Singapore or the Indian states or Spain or Italy, we had looked at houses, planning always to stay there for a time. Perhaps we would have stayed but for children and dogs and education and such domestic matters. There had always to be a base, a center where there was a permanent establishment for children and dogs, and the vast accumulation of pictures, books and souvenirs which kept growing despite anything one could do. Once when Isabel Paterson was asked how she was going to furnish a new house she had acquired, she replied, "I've never had any trouble about that. The only trouble I've ever had was to keep furniture out of the house."

That too had always been our experience. Wherever we have lived, no matter how big the house, the attic, the outbuildings, sooner or later became crowded with extra chairs, books, pictures and toys.

The nearest to a permanent home we had ever had was the house in France and that has never really belonged to us because we could not buy it. We could only lease it for fifty years. A half century, it is true, is a long time. That old eighteenth century house was our home and base for fifteen years and we spent a great deal of time and money and work upon it until it was in all but deed completely ours. But still there hovered over the whole adventure the dread sense of impermanence and doom—that one day there would be war and perhaps the German cavalry would again quarter its horses in the ancient chapel and as they had done in 1870 and in 1914 carry off the furniture, the pictures and books into Germany. And about that house, much as we loved it, there was no sense of real permanence, no feeling that one day grandchildren and great-grandchildren would return there to ride the pony and fish and go swimming in the little river. There was always about it a sense of living upon borrowed time in a dream which one day would come abruptly to an end. During all those years there was in every one of us, even the smallest of the children, I think, the hunger for a house of our own outside the doomed world of Europe.

In the end Mary and the children left that lovely old house as refugees. Most of the books, furniture and pictures were shipped out to safety a little while before the Germans came in. Some of them had real value in terms of a market, but all of them had rich sentimental values. Some were the gifts of friends, others of artists and writers and people of every sort and nationality. Most of the furniture was beautiful French furniture touched with that sense of immortality and permanence which the French of all people put into the things they make and build. Most of that furniture, made by fine craftsmen of the eighteenth and nineteenth centuries in France, is sheltered now in a big farmhouse in the middle of Ohio, a better fate for it, I believe, than to find itself in some middle-class house in Stuttgart or Leipzig. The love, the care, that those craftsmen, now long dead, lavished upon the chairs and tables and bookcases is still alive in a country five thousand miles away, a country which, oddly enough, was itself a wilderness claimed by the French at the time the workmen bent over their lathes and chisels carving the traditional designs of Brittany and the Valois into the beautiful walnut and pearwood.

Since the house was to shelter us as a family for the rest of our existence and perhaps to be passed on to grandchildren and great-grandchildren many things had to be considered. The same sense of uncertainty of the future which had given me the hunger for self-sufficiency and independence played a great part in the plans. The house had to be big so that it could always be a refuge for children and grandchildren and even for remote relatives and friends.

The sense of responsibility was strong in both my wife and myself for it had fallen to both of us, not always to our pleasure, to take care of the indigent and black-sheep members of our own respective families even to such remote relationships as third cousins twice removed. It was handed down to us in turn from our respective parents. All through my childhood, our house was always filled with people temporarily down and out. Remote relatives came to my mother's house when they were penniless or ill; they came there to die and be buried from our house. They were married there. Some were conceived and born there. At times whole families descended upon us during periods of ill luck. My childhood memories were filled with a sense of being crowded, of living in a moderate-sized house too filled with people, with strong and violent personalities always in conflict. It is not a good way to live.

And so, when at last it came to building a house in which we were to live for the rest of our lives, both my wife and myself were under the compulsion, born of childhood memories, of creating a house big enough to shelter all kinds of people and still provide us with a reasonable degree of privacy and human dignity. I earned much money by writing and in the background there was always Hollywood when the money ran short. As George once suggested, there should be plaques placed on each of the farm buildings announcing that "Twentieth Century Fox is responsible for the building of this sheep barn," or "Metro Goldwyn Mayer provided the money for remodeling this cattle-feeding barn." "United Artists, in payment for a short story, built this cottage."

The name for the whole farm came naturally. The bulk of the capital which went into the whole adventure came from books written about India, so it seemed only fitting that the farm should have an Indian name. We settled at last for Malabar, the name borne by the beautiful hill overlooking the harbor of Bombay and the name also of the southwest coast of India, one of the most beautiful parts of the earth, where we had lived happily for several months. Malabar Farm it became.

But the house had to be more than merely an asylum for indigent relatives and friends. It had also to be a kind of hotel where people with whom I had business dealings could visit. I had always preferred to stay at home for dinner rather than to go out, to have people come to my house rather than to go to theirs. And now in the hills in Ohio, it was not merely a question of going out to lunch or dinner; it was a question of going to New York and Washington and Los Angeles and Chicago and other places where there was business to be done. And so there had to be rooms in the new house for business agents, for political and literary people, for actors and actresses, for visiting foreigners, for foresters and farmers and professors and just plain friends.

Even in the small house in France there were always people staying with us and when there was no more room in the house, the overflow stayed at the inn called Le Grand Cerf, a little way from our house. Sometimes, on Sundays, as many as fifty people came to lunch. They came from all parts of the world, every sort of person, of every race, creed and color. All that too was in the blood. Our grandparents, our parents, had lived that way and we meant to go on living that way for the rest of our lives, rich or poor, in a big house or a small one. So long as there was money, it would be a big house in which our friends would be comfortable.

I knew, shrewdly I think, that the only way anyone could any longer lead that kind of life was in the country or on a farm where one had one's own pork and beef, milk and eggs, vegetables and fruit, even one's honey and maple syrup. It had to be a world in which there was plenty, which ration books and scarcities such as I had known in Europe could never touch.

And beyond all that was a curious sense of obligation shared by all of us, almost a kind of superstition, I think, that we should share our good fortune in life with others. We had lived thus in France among our French neighbors and friends and we wanted to live that way in the midst of Ohio, with a house which was also shared by all our neighbors and friends. That too called for a big house where dances could be held and meetings and picnics and reunions.

. . .

Altogether it was a big plan, requiring money, energy, faith, knowledge and enthusiasm, through fortunate circumstances and some intelligent planning none of these were altogether lacking. Perhaps the most important rule we made was that the undertaking was to be carried out upon a reasonable basis without the expenditure of large sums of money. It was neither to be a plush and gold-fitted establishment of the sort in which "city farmers" once indulged, nor the kind of impossibly conceived cooperative farm which had been launched with government money. Both Max and I were practical enough to know that such a subsidized venture proved nothing and did less to help solve the sickness of agriculture. Nor did we believe that such cooperatives would ever succeed either economically or on the human basis. We determined in the beginning to do nothing which the average farmer could not afford to do. Whatever experiment we made, whatever building we did, whatever restoration we undertook would be in the terms of ordinary farming and within the economic possibilities of any family-sized farm enterprise.

We sought to prove that run-down land which had become virtually an economic liability to the nation could be turned into an eco nomic asset and that farming, done properly and managed intelligently, could provide an excellent investment for capital. We were aware of two advantages over many small and all submarginal farmers. Backed by the proper capital, we should never be forced to sell in a bad market in order to pay interest or taxes. We could, often enough, be able to deal with the ruinous middleman on our terms and not his.

I think what all of us were trying really to find was the old spacious comfortable life which farmers and landowners once lived. We believed that it was possible to live that way again—the best life, I believe, that man can know—if intelligence and skill and hard work were employed in building and carrying out the project. We were building on a fairly large scale, but scale, we believed, had nothing to do with it. A man could have that good kind of life whether he had fifty acres or fifty thousand. Whether we failed or succeeded was important to us, but not so important as the adventure itself.

None of us held the illusion that in order to be a "dirt farmer" we had to be louts who lived in the squalid conditions of "Tobacco Road." That illusion, that legend, was in itself a sign of the sickness of American agriculture. Max's father had not lived like that, nor my grandfather, nor the greatest of farmers and Americans—Thomas Jefferson. We all believed that in the soil could we find not only security but the best life in the world.

THE BIG HOUSE

(A chapter to be skipped by those who have no interest in architecture.)

In a way, although there were plans and blueprints, the Big House was built the way a house should be built, bit by bit as we went along. A house must, like the soil, be a living thing or it is nothing at all but walls and roof and cellar.

In novels and in stories I have often written of houses—houses that were bleak and cold and without warmth or color, houses that were so ancient that in them was recorded the history of the very land on which they stood, houses which were old and shabby but beautiful because they had been lived in, houses which were new and bright and modern and chill as icebergs, houses that were gay and houses that weighed like lead upon the spirits. I have an indestructible conviction that houses become in time stamped with the character of their owners. There is a kind of aura about every house I have ever entered, so strong that I believe I could tell you a great deal about the owners after ten minutes spent within the walls—whether the wife was dominant, whether the family was happy or unhappy, and almost exactly the degree of education and culture and knowledge of the person who built and furnished and lived in it. In short for me there are few things in life more interesting and revealing than the houses in which people live.

We have lived in houses in England and in France and America and spent long periods in houses in Spain, in Austria, in Switzerland, in Sweden, in India and elsewhere and all of that experience was good training for the day when we came to create—create rather than build—

the house in which we, as a family, expected to spend the rest of our lives. We knew, by the time the family was half grown, pretty well what sort of a house we needed. It was not simply the problem of putting up four walls and a roof as a shelter, nor a problem which could be solved simply by asking an architect to work it out and deliver it all finished and ready to occupy. Houses affect the lives and character and happiness of people who live in them as much as all these things affect the houses themselves. I know of houses which have caused divorces and deformed the lives of children growing up in them, because they were badly planned for the personalities of the people who occupied them. I know that almost any reader who has lived in many houses has had the experience of hating certain houses, partly because of the aura left by predecessors and partly because of the stupidity or harshness of the house itself.

We had countless elements to consider. One thing was that we were a family of *big* people, big physically and rather big and loose and careless in our living. We had to have room for physique, for personality and for spirit. None of us, including George and Nanny, down to Ellen the youngest could be said to have a dovelike submissive personality. I doubt that any family was individually ever more decided and stubborn, in opinion and actions. It was not simply the stubbornness and pigheadedness which concerns personal relationships. It ran much deeper and much wider to questions of politics and international affairs. (During the closing days of the 1944 political campaign when the house was divided into two hostile camps, the opposing factions, save at mealtimes, virtually occupied two opposite ends of the house.)

And we had strong beliefs in the civilization of French and English middle-class households where the children and their friends were able to lead their own lives, apart, without overrunning the adult part of the family and without being crowded or dominated by grownups. Children and adults living in too close proximity can cripple each other's personalities and make scars which remain, especially with the children, for the rest of their lives.

And there was the question of convenience. All of us had lived together long enough to know what each of us needed—Nanny and the children plenty of room and rooms which were built to take hard wear and tear and be easy to clean. George needed a kind of hideaway, neatly kept, where he could do his work and entertain his friends and, when he saw fit, lead a life apart for days at a time. Mary needed a kind of feminine nest with considerable *froufrou* and disorder, and I myself needed a big rambling room with a lot of light and a lot of big comfortable chairs and sofas on which a tall man could lie on the base of his spine and where you could talk for hours undisturbed with friends about farming or politics or international affairs. I needed plenty of wall space for books and plenty of space for a giant desk and room for five or six dogs. It had to be a complete unit for living a life apart, a big room in which to sleep, to work, to read, and on occasion to rest. There had to be room for seeds and farm books and pamphlets, for both country and city clothes, for the pruning shears and saws and trowels that were my personal property, and room for the vast accumulation of photographs, souvenirs, posters, letters, etc. which I collect like a crow wherever I stay for a little time. (Yesterday we carted out from the room four wheelbarrow loads of books, papers, letters, photographs and old clothes which I relinquished at last only because they were crowding me out of the room.) The room had to be a bedroom, a farm office, a study, a workroom, a sitting room. It had to be built for someone who was likely to work at any hour of the day or night.

But there were other problems to be considered—the fact that the house was also to be a kind of center for the community in which dances, reunions, meetings could be held, and the fact that the very site, a long shelf halfway up a long hill, with a steep-sloping old orchard and a steep outcrop of sandstone just behind it, demanded a long low rambling house. That problem was solved by the desire of the whole family for a house that was like a French house of the eighteenth century, with windows on both sides of every room and doors opening from every room on the ground floor into the garden. We are a family which lives out-of-doors a large part of the year more than it lives indoors, a family which will always sit on a terrace or on the grass until a blizzard drives us indoors. And the dogs had to have plenty of doors.

Most of all, perhaps, I wanted a house which after a year or two looked as if it belonged there on that hillside shelf in the middle of the rich Ohio country, a house that looked as if it had been there since the clearing of the wilderness, above all, a big house which would not stick up like a sore thumb in the midst of a beautiful landscape.

With all these problems in mind, and many others, we discussed the project with Louis Lamoreux, an architect of good taste who knew his Ohio countryside and his Ohio architecture, for he had once made a survey and report on old houses in Ohio. I did say to him in the very beginning that he would have a tough job in which, at times, he must be willing to become draftsman rather than architect, and that very likely he would have to tear down and rebuild as he went along. He was willing to take on the task and found, I know, that I had not exaggerated the difficulties and complications.

Because the Greek revival style of Thomas Jefferson had left a great imprint on all early houses in Ohio, I wanted that style to dominate both the outside and inside. There were many reasons for this: (1) because the style belonged in Ohio; (2) because we had many French pictures and much French furniture rescued at the last moment from the Germans and all French things fitted into the frame of the classical Greek revival style; (3) because it seemed to me there are few architectural styles more right and more beautiful for farmhouses and houses in small towns where there are open lawns and shrubs and great trees.

The Greek revival style dominates all old buildings in southern Ohio, but in the northern part, in the Western Reserve and the Firelands, the old houses are copied from the English Georgian style, filtered through old New England towns and adapted by frontier workmen from old houses in Salem and Newburyport and Ipswich and Portsmouth. In Ohio, Malabar lies almost exactly on the border line between the two styles. North of us there are towns like Norwalk which appears with its white New England houses and big elms, like a town in Massachusetts or Vermont, and south of us there are old towns like Mount Vernon, and Granville that are like the loveliest of Southern towns, although they are free from the decay and shabbiness that afflicts so many old towns south of the Mason-Dixon line. I wanted in the Big House some recognition of the New England style of northern Ohio.

To help us there was the friendly knowledge and advice of I.T. Frary, of the Cleveland Museum, a friend of the architect, who knew his Ohio houses to the ground and had written many books on the subject. He was, too, perhaps the leading authority in the country on the Jefferson Greek revival style. The idea of building a house which was to be a kind of apotheosis of Ohio architecture appealed to him as it appealed to Louis Lamoreux, the architect.

We lived then in the big house on the Fleming place built from a mail-order house plan which, save for practical aspects, was everything that an Ohio farmhouse should not be. Scale and proportion had never existed in the mind of the draftsman who conceived it. Everything about it was wrong, its gables, its windows, its doors, its porches. All were monstrosities, doubly noticeable and offensive in a valley where the old farmhouses were beautiful. It had been conceived, like nearly all houses at the turn of the century, not with taste or knowledge or any regard for beauty, but to show off the success of the owner. Fortunately it is virtually concealed by big elms and maples during most of the year.

In this monstrosity, Lamoreux and Frary and the family spent many hours working out the details of the Big House.

There stood on the site of the old Anson place, a plain square, uncompromising house without architectural detail but possessing the beauty of an honest, functional house, honestly conceived and honestly built. At times we considered doing away with it altogether and starting afresh but always at the last moment I could not bring myself to break the line of continuity and tradition which has always had a strong appeal for me. I do not believe that any style either in architecture or decoration or literature or music which breaks clean with the styles and traditions of the past has any chance of survival. Somewhere there must be the discernible line which connects the present with the past, the new with the old. The presence of that line in Dutch and Swedish modern architecture and decoration has, I believe, determined its survival and acceptance while the French and German modern styles have already become outdated, abandoned curiosities. The French and German styles did not grow out of the rich past but sought only to break with *all* tradition and create forms and styles which were unlike anything ever seen before and they produced monstrosities such as chairs which were unsittable, bookcases which hung suspended from the ceiling and required a ladder to reach the books, and interiors which were more like an operating amphitheater than a room to be lived in.

Styles in architecture, even elaborate styles, have grown up through centuries to suit themselves to climates and living conditions of the countries in which they developed. They belong essentially in these climates and conditions since they grew out of them. That is why a Spanish house with a patio is an absurdity in Long Island and a Swedish baroque house or a Tudor house is a monstrosity in Southern California.

And so it was decided that the Anson house should remain as the nucleus of the new house and that additions totaling approximately twice the space of the Anson house should be added. The new additions were to be built at different levels, with slightly varying styles so that they would give the impression of having been added from time to time over a period of a hundred years or more.

There were certain features of the Anson house which we wished especially to keep. One was the old cellar with its walls of pink and red sandstone crudely dressed by the workmen who built the house. The other was the twin front doors which were characteristic of the old houses built by the early Pennsylvania settlers in our Valley. Most of the houses have such doors, two of them side by side. One led into the living room and one into the parlor. The one leading into the parlor was opened only on the occasion of epochal ceremonies— a christening, a marriage, or a funeral; the other leading directly into the living room served as the main entrance to the house. Those two doors eventually came to open, one into the farm sitting room and the other into Mary's ground floor bedroom. The old tradition has not been kept and both are used today all through the summer and winter months for they open onto one of the two big verandas.

One thing troubled me and that was the question of fine workmanship. In Europe craftsmen still existed and the man who worked by hand took intense personal pride in each piece of furniture or woodwork or ornamentation. The result had none of the mechanical deadness of the machine-tooled job. There had been a time in Europe and in America too when a chair was considered not merely as something to sit on but a work of art, worthy of bearing the trademark or the signature of its creator. We had such pieces of furniture brought from France, and once you have lived among such furniture and woodwork, the machine-made kind often seems soulless and even ugly.

Architectural details were of as much importance to Lamoreux and Frary as to myself and they seemed less troubled than myself about the quality of American workmanship. In the end the interior detail and some of the exterior work was done by a wood-working company in Alliance, a town sixty miles away, which did a remarkable and beautiful job. The carpentering was done by a firm of three brothers called Bauer and their helpers who were as fine craftsmen as Europe can produce. They were Hungarians of remote Austrian origin and pride in workmanship was not lacking. For the furniture and fine cabinetmaking we found an aged German cabinetmaker called Schlitter living in the county seat of Mansfield who worked with as great skill, pride and taste as any *menuisier* of the great days of France. In his shop he kept great stores of fine woods, well aged and seasoned—walnut, pearwood, cherry and all the other fine hardwoods. He bought old pianos and the old cherry and black walnut beams from old farmhouses and barns that were being remodeled or torn down. Later when the French furniture

showed signs of cracking or falling apart from the drier climate and the central heating he remade or repaired most of it with the respect of one artist for the work of another. Lester Bauer, one of the three Bauer brothers, was one of the two carpenters left in America who could still build a self-supporting circular stairway and he was called to all parts of the country when such stairways had to be built. But both the Bauers and Mr. Schlitter still had roots in Europe and the traditions which made for fine furniture and woodwork were still alive in them. Schlitter was born and learned his art in the Rhineland and the Bauers were the first generation in America descended from fine Hungarian and Austrian workmen. I doubt that they will have successors in a country where standardized machine-made copies at bargain prices are the goal.

As a result of their efforts, the Big House had, when it was finished, that atmosphere of personal, individual craftsmanship which houses in Europe and old houses in America have. If it is still standing a couple of hundred years hence, architects and connoisseurs will believe, I think, that the house was built not in 1939 but at least a hundred years earlier.

The building and decorating took nearly eighteen months, partly because of an exceedingly cold winter and partly because it was built with deliberate slowness, piece by piece, with great care that each piece should be exactly right. During that period I had to be away much of the time and when I returned found often enough that something had gone wrong or that the result was not as I had visualized it. More often than not this was the fault not of the architect or the builders but of myself. I possess, unfortunately, a hopelessly unmechanical mind with an utter inability to understand blueprints or to visualize from blueprints the final result. Louis Lamoreux had never before built what might be called a French house with the classical sense of scale and balance and space which all French houses of the eighteenth century possess, and he was inclined to overlook these elements. The result was that very often on my return, walls had to be torn down, partitions taken out, the scale of doors enlarged and the pitch of steps altered. This was not an economical way to build a house but it was, I think, a wise one and cheaper than making the changes at a later date when the house was finished. I knew that we were a family who could not live happily in a house in which balance, scale and space were not scrupulously observed, and this house was being built to live in for the rest of our lives.

And so as we built and rebuilt, the house began to emerge. I know of no greater satisfaction than the construction of a house, especially after the fashion in which we built the Big House. Gradually the story of it, purely fictional, began to emerge. It began with the Anson house, a square, uncompromising structure of good proportions and scale. Then (the story ran) the first farmer during a period of prosperity had added the west wing which included the two big workrooms occupied by myself and George, mine on the ground floor, George's on the second floor. Then at a later period, perhaps two generations later, as the family grew, the first part of the north wing was added, in a little different style with different materials, and finally about 1880 the kitchen wing was added to the north wing. It was built simply with utility the principal goal.

That was the fiction and although all save the original Anson house was built at the same time, we began to see the house in that fashion and all of us, family, architect, carpenters and stonemasons came in a way to believe the story. The fiction succeeded for even when the house was first finished it gave the impression from the exterior that it had grown over a period of a century rather than having been built all at once. That effect was even mere true of the interior with the different levels and the steps leading up or down from the different parts of the house.

As the house took form, the details which Lamoreux and Frary had contributed began to emerge. There were the two big porticoes with cast-iron pillars in a design of grapes and leaves which came from old houses along the Ohio River. There was a wooden portico over the main door intricate in design but beautiful, copied from a house at Norwalk, and over it on the main cupola a simple but beautiful fanlight which came from a house in the original Mormon settlement at Kirtland and may have been born in the head of Brigham Young himself. Brigham was a carpenter and a good workman and had an undisciplined passion for architecture which did not always achieve beautiful or harmonious results. A good many of his creations included what he considered the best points of Gothic, Moorish, Greek, Renaissance and baroque architecture, all employed in a single structure. I know no better example than Brigham of the danger of enthusiasm without the taste and understanding which can only come from knowledge and understanding and the discipline of the classic tradition. But the simple classic fanlight is all of a piece and a fine ornament for any country house.

The dormer windows of the north wing came from the old inn at Zoar, built in the heyday of the Zoarite communal settlement there. They are simple, direct and in good scale. The little interior portico leading to the front door and the big hospitable door itself, so characteristic of the New England houses of the Western Reserve and the Firelands, was copied intact from an old house at Chagrin Falls which had long since burned down. The railings of the upper porches, very simple and direct, were copies from the inspiration of some long-dead, unknown carpenter craftsman who worked on a farmhouse near Danville. The great chimneys made of the big red and cream-colored blocks of sandstone from the foundations of the old vanished mills and the foundations of the burned house on the Ferguson place, were like the great chimneys of the taverns in the Western Reserve. The ornamentation of the doorways and windows came from the famous Taft-Sinton house in Cincinnati.

The house spread, long and low, on its shelf along the hillside fitting in among the trees that already stood there—two immense Norway spruce, an ancient black walnut, a big catalpa and a great cluster of balm of Gilead poplars just below and beside the new pond that was beginning to fill with spring water. From the distant Valley Road the house appeared to have grown there on that shelf of land to be as old as the trees themselves or as if, with great skill, the architect had constructed the house as one piece and then slid it into place beneath the trees. Clearly, it wasn't only a house; it was, despite its size, a home; what the French call a *foyer*, lived in and made for living. It was all Ohio and it was of a piece with the great barn with its cupolas and ornate shutters.

Inside the house the same source of details played a part. Throughout the house in moldings, in the big black walnut doors so typical of Ohio, in cornices, there are copied designs which are tributes to the taste and craftsmanship of men who worked long ago with the native woods and the simple tools of the frontier. The cornices, moldings and doors were in turn copies of decoration and architecture in the seaboard states and in England itself. Some of them, touched by the simplicity and directness of the frontier, have a beauty surpassing that of the originals.

The big hall, with its twin stairways and the big doorways at either end is the purest Jefferson Greek-revival in style with the simplicity and dignity which touched everything with which our most civilized American concerned himself. At the turn of the stairs there are two niches built to contain the Houdon busts in plaster of Jefferson and Lafayette. It is a hall big enough for music and for dances and a couple of times a year the Lucas High School band of forty-two pieces plays there noisily at the meetings of the Farm Bureau and the Alumni Association.

The big living room is as nearly the classic proportion of two cubes as the limitation of site and scale permitted, with four big French doors leading to a veranda on one side and directly into the garden on the other. It is a French room with yellow-gray walls and a mirrored wall into which is set a great mantelpiece from France by way of New Orleans. Over it, on the mirror is a big gilt American eagle surrounded by gilt stars of the forty-eight states, and opposite in a niche set among the books is the Houdon bust of Voltaire. It isn't a parlor. It is a big and cheerful room about which centers the whole life of the house and the farm. With rugs and chairs and sofas removed, it serves two or three times a year as a ballroom.

The dining room, like the big living room, is French, of simple Louis Quinze baroque and at each end there are two deep bay windows running down to the floor which give out on one side over the whole of Pleasant Valley and on the other toward the old orchard and the jutting outcrop of pink sandstone behind it. Both rooms are vague nostalgic copies of similar rooms in the house in France wrecked by the Germans.

There are two dining rooms, a second known as the "country dining room" for the children on occasion and for the people who come and go, working at various tasks and for the boys interested in agriculture who come each year to work at Malabar during the summer holiday. Both dining rooms are nearly always filled, for in summer we start off each day feeding twenty people, not counting guests and visitors. Nevertheless there is always an extra place or two at the tables for unexpected arrivals.

There are other bay windows in the house besides those in the dining room. In fact there are bay windows and balconies wherever it was possible to have them without marring the architecture of the exterior. For me bay windows seem to bring the out-of-doors into the house, and they give much light during the long gray days of winter. I had always liked them in English houses but it was in Sweden that I saw them used to greatest advantage in a climate where the days in winter are short but usually bright with sunlight. They fit undisturbingly into the baroque style in which so many Swedish houses are built and were used wonderfully in a house in Stockholm belonging to the elder Bonnier of the Great Swedish publishing firm of Bonnier. In that house nearly every window was a bay window, some modified and shallow, but all serving to bring light into the house, and in that house they were kept filled with potted flowering plants, winter and summer, so that when you entered the house and looked down the long succession of rooms, you had the impression, even in winter, of entering a garden. In the Big House we followed the Bonnier plan of bay windows filled with plants.

And there was another plan used with regard to windows which was the idea of Lamoreux, the architect. All the windows in the house, and naturally the French doors, are set low so that one can sit in a chair or even lie in bed and look out over the landscape. Never is anyone forced to stand up to look out of a window. The space beneath each window is used architecturally for books or shallow cupboards to hide the things collected by a family with the instincts of pack rats, or they serve to hide the hot water radiators concealed behind what are no more or less than simple adaptations of the shutters on the outside of the house, painted to match the room. I have a congenital dislike for the appearance of radios and radiators. The radio problem was solved by placing a radio in the closet beneath one of the stairways with wired in, concealed loudspeakers in the principal living rooms which can be turned on and off at will.

At the east end of the house is the "country" living room, a good, hard-worn, sitting room with a table for drinks and ice.

Fortunately everybody in the family liked color, not the muddy colors that were neither green nor blue nor red, but clean colors that were clear and gay, and so the interior of the house when finished, was cheerful and bright with none of the dun-colored dreariness of so many houses in which the owners are afraid of colors. There was lots of yellow and red and green. The dining room was the baroque green that one finds in houses in Austria, the big living room yellow and gray, the Greek revival hall white and gold, the little farm sitting room, lacquer red and white with a country wallpaper with a design of abundance—red raspberries spilling out of white shells. My own room was Paris green, the bright acid green of the painter Veronese. The colors made you feel good as you walked through the house—the colors and the books which fill in wall spaces in almost every room.

The Anson cellar with its old dressed sandstone walls became a meeting place that will hold about two hundred people, with a big fireplace at one end, and the room next to it became a billiard room with two secondhand tables much used by all the farm and by the neighbors in the long winter evenings.

Every member of the family planned and decorated his own room to suit his needs and taste. The result has been both variety and satisfaction from my own big general utility room on the ground floor to the children's rooms with the dormer windows copied from the old inn at Zoar. There wasn't any decorator or really any single architect for the house. It grew, with ideas and suggestions made by friends who were painters or architects or decorators, some of them famous in their respective fields, who came to Malabar while the Big House was being built or afterward.

Architecturally speaking, some trickery was employed to achieve the interior effect of height. We were faced with the problem of having high-ceilinged rooms in a house built upon a site which demanded a long, low house. And so when the high ceilings suited to a tall family became impossible, the only refuge lay in illusion. It meant the use of vertical lines rather than horizontal lines. The French doors and the big interior doors of black walnut rise from the floor all the way to the ceiling. The curtains hang all the way from the ceiling cornice to the floor. The rail of the double stairway in the hall was made three inches lower than the conventional stairway to give the effect of increased height. By and large the effect of a foot or two of extra height was achieved in each room.

There is a fashion in these times to deprecate good, solid construction, to be ashamed of background and knowledge, luxury and even comfort. It is the fashion to believe that a dirt farmer must belong on Tobacco Road, that one should live not only modestly but inconveniently, in patterned houses and flats built for housing by the government. It is a part of a growing philosophy of deprecating individuality, of dragging all things down to the level of the lazy, the uninspired, the shiftless, the dull. It is a philosophy to which I do not subscribe and never expect to until I quit this world. Houses should be the expression of the people who live in them. They should belong to people, not to communities—a fact which the Swedes recognize so wisely in their government building projects. Any other philosophy is a nullification of the very meaning of civilization. Also it is an unnatural philosophy and therefore doomed forever to eventual sterility.

Our house is a big house, well built, to be used not only by ourselves but by friends and neighbors as well, and by generations after we are dead. Already it has had hard use which has mellowed it and made it shabby. There is no parlor and no spare room kept empty with the shades drawn. Every inch of it has been in hard use since it was built and will, I hope, go on being used in the same fashion so long as it stands. Perhaps one day it will belong to the State, together with the hills, valleys and woods of Malabar Farm. It cost money but it was money that all of us earned and although all of us, even the children, I think, got great pleasure out of planning and building it, we haven't kept the pleasure of using it altogether to ourselves. It stands overlooking the Valley which I loved as a boy and still love better than any valley in the world. I fished the creek below and ranged the hills and woods hunting and gathering nuts as a small boy. I swam in the swimming hole where now the kids of the farm go swimming. I find there the continuity which existed in France, that growing of one thing into another, the succession of generation by generation, which is the rich, satisfying rule of Nature herself and indeed of all civilization. I think the dream of that house was there long ago in the days when as a small boy I knew every tree and spring and pasture in the Valley.

ON BEING *TECHED*

THERE ONCE LIVED IN OUR COUNTY A REMARKABLE WOMAN whose name was Phoebe Wise. Vaguely she was a relative, her grandfather and one of my paternal great-grandfathers having been brothers. I have written of her many times although while she still lived I called her by other names than her own. It was useless to attempt to disguise her character and personality by using an invented name. No one who knew her was deceived by the strategy and I am told that she, herself, was the least deceived of any, and that the stories I wrote about her caused her pleasure and even mirth, since the invented portions were, she considered, entirely outside her character, especially on the romantic side. For Phoebe, it must be said, had contempt for nearly all men, to the point of actually having murdered one, half intentionally. Indeed, I have written so frequently of her that at times I have very nearly worn out the subject, like an etching plate which has grown old and worn from use. Yet I cannot stop myself even now, for I think I never met any individual who left upon me so profound an impression. In all the years I was away from the Valley I never forgot her. The first time I saw her I was about seven years old and the last time I saw her I was seventeen. She died at the age of ninety-three while I was on the other side of the world.

My memory of her is of a tall woman with a remarkably fine figure and carriage even at the age of seventy when I saw her last. At that time her hair was still black as a crow's wing. I think it stayed black because of her remote Indian blood. Her grandfather, my great-grandfather's brother, who was a circuit-riding preacher in the days of the first settlements, married the daughter of a Delaware Indian chieftain who was one of his converts, and all her life Phoebe bore the stamp of that remote Indian blood, in her carriage, in the blackness of her hair and in the fierce blackness of her eyes.

The last time I saw her I was home for a few days before leaving for France and the war of 1914 and she stopped to talk with my father and me, to wish me good-bye and good luck. She was dressed, as she always was, in an extraordinary fashion. She wore a yellow taffeta dress with a bustle and many flounces, very long with a slight train which she did not lift but permitted to drag grandly behind her. Weather, mud and dust had discolored the yellow taffeta to about the height of her knees. On her hands she wore black lace mitts and her fingers glittered with cheap rings in which were set imitation diamonds, emeralds and rubies. On her head she wore a large rusty black picture hat, and on it for ornament, she had pinned a garland of real wild asters plucked from the roadside. For as far back as I or anyone else could remember she had worn the same hat, changing the fresh flowers each time she came into town, according to the flowers in season. During the winter she pinned sprays of wheat or oats to the hat. Over one arm she carried a basket in which were collected the purchases she had made on her trip into town from her house which stood a mile or two from the factories in the shadow of the state reformatory.

It was a tiny house with a small turret and a great many tiny gables and much fretwork, overgrown, rotting and forced apart by the thrusting shoots of wisteria and trumpet vine. The little garden which surrounded it was a jungle of old-fashioned flowers—lilacs, mock orange, day lilies, petunias and a hundred other shrubs and plants. Here she lived alone surrounded by all sorts of animals, both wild and tamed. Between them and her there existed an extraordinary relationship. She allowed no hunters on or near her place and she had been known to drive them away on more than one occasion with an old-fashioned musket. Although it was doubtful whether the rusty musket could have done anyone the least harm, she carried the prestige of once having shot and killed a man, and trespassers did not pause to argue with her. Her only guardian was not a dog but an old white horse which she had raised from a colt without once having harnessed or bridled it. The old horse would run at trespassers, showing his teeth and kicking out viciously with his forefeet. You could not enter the place unless she called off the horse which obeyed her exactly as if it had been a watchdog.

From the description I have given you may gather that her appearance was that of a dowdy scarecrow. Actually the effect was exactly the opposite. In the first place the old-fashioned clothes had about them an air of great style, even the real flowers were pinned on the old hat with a sense of style that would have brought credit to a great Paris milliner. And she was one of those women who, by her carriage and the thrust of her head, could invest even a Mother Hubbard with style. Beyond that she always had an air of immense dignity, almost of majesty which, together with her strange but clear brain, not only demanded but commanded the respect of all who knew her. In the county Phoebe was known as a "smart woman" and what they meant I think was that she was remarkably well informed, that her brain really functioned and that she was quite able to care for herself. On the last day I ever saw her she talked of Europe and the war there with astonishing clarity, perception and knowledge.

The story of the man she killed was a simple enough one. He was an unwelcome suitor, one of the many who sought to marry her when she was left an orphan with a small inheritance at the age of eighteen. At that time she lived alone in the same little house without even an old white horse as a guardian. The ill-fated lover was unwilling to accept her rejection of him and on more than one occasion had come to the house late at night and attempted to force his way in. At last one night, her patience gone, she called through the door that if he did not go away, she would fire through the door. He made one more attempt to enter and she fired. She heard no more sounds from him that night, but in the morning when she opened the door, she found him outside on the doorstep, dead.

At the trial Phoebe insisted upon defending herself and was, of course, acquitted, but the knowledge that she had killed a man set her somewhat aside from the other members of her community and undoubtedly strengthened her determination never to marry. From then on, although she walked to town once or twice a week, her life became more and more solitary and her natural intimacy with birds and animals increased.

The odd thing was that she really liked society and intelligent conversation, but she satisfied the liking, not in her own little house to which few people ever were admitted, but on street corners where she would stand, in her fantastic clothes, talking for hours with the men and women whom she liked or respected. She had a sympathy for the simple ones and the eccentric characters like herself. My father she deigned to talk with because he was a simple and friendly man and vaguely a relation. She always called him "Cousin Charley." Usually you found her talking with Gus Douglass, a brilliant but eccentric lawyer whose own sense of mockery and humor destroyed his respect for legal processes and blighted what might have been a great but a far less human and satisfactory career. Or it might be Susie Sturges, another eccentric as strange in dress and appearance as Phoebe herself, or anyone who she felt was *free* or simple and direct or intelligent. She was immensely selective—Phoebe—and in her own way, a snob. She had no time for the conservative or the conventional whom she held in contempt as limpets clinging to the rock of conformity.

I have written at length about Phoebe because the memory of her has always been so much a part of my own life and because, her legend, like that of Johnny Appleseed, has persisted and grown as a part of the life of that beautiful country. But there is another reason, perhaps more profound, which has a bearing upon this story. The memory of it is still vivid, although the observation was made by Phoebe when I was about seven years old.

I had gone with my father electioneering and we stopped at Phoebe's little house out of sheer friendliness because Phoebe, being a kind of Rousseau anarchist, held all politics in contempt and did not vote. While my father talked with her I played in the jungly garden which surrounded the house, making friends with the animals, both wild and tame, which lived there. Among them, I remember, was the white horse which at that time was only a colt perhaps not more than a year old. I don't know what kept my father talking with Phoebe for so long but presently I was aware that they were both coming down the path toward the springhouse, Phoebe dressed in the practical man's clothing which she always wore around her own place.

I was sailing boats made of twigs on the surface of the spring pond. The white colt was standing beside me, playfully dipping his muzzle into the cold water and then raising his head and tossing it so that the water fell in a shower over me. It was a comic, clownish trick, one of those which at times convinces me that animals know far more than we suspect, that some among them possess even a sense of humor. I was absorbed in my play and did not look up but I heard Phoebe saying, "Cousin Charley, that boy is *teched* too."

As I turned, I saw her watching me with her burning black eyes and suddenly there passed between us a strange current of sympathy and knowledge, which bound us together forever afterward. Because of that look I remembered her in strange, far-off places. Sometimes I dreamed of her. Because of that look I was glad that on the day I left for the war, the unearthly, witchlike Phoebe was almost the last person I saw on a street corner before Ashbrook's drugstore. When she wished me good luck, I had a feeling that nothing would happen to me. Curiously the feeling persisted throughout the war even under circumstances in which it appeared unlikely that any man in our company could ever come through alive.

Afterward, as we were driving away from Phoebe's in the buggy I asked my father what "teched" meant. He laughed and replied, "She means a little crazy, like herself." And then after a little time he added with a sigh, "A lot of people think Phoebe is crazy, but I don't. I think she's awful smart."

It took me a good many years to understand to the full what Phoebe meant by being "teched." It needed a lot of experience and a lot of observation but I think I know now what it was that lay behind the remark. I know today that any good farmer has to be a little "teched" and when I go over the list of good farmers I know, there is not one of whom it could not be said, "He is a little 'teched,'" for it means that he loves his land, his animals and his trees and understands them all. He farms not in order to make money but because of the pleasure and the satisfaction there is in it, because it is a destiny he would not change for any other in the world. Success and profit follow, but they are merely incidental,

And that, I think, is a rule that holds through all of life; the ones who set out in life with the sole object of "getting rich" rarely succeed in their ambitions, and as a rule they lead narrow, pinched hard lives unilluminated by the warmth of idea or ideal. They remain shallow egotists, isolated both from man and Nature. The man who has set money as his only goal and failed is the bitterest and most perverted of all specimens of humanity.

I think all that was a part of Phoebe's strange wisdom, and now, when I am a middle-aged man and Phoebe is in her grave and I have acquired a humble stock of experience and wisdom and humility, I know that the casual observation made beside the spring is the finest praise I have ever had. And I know that all the real satisfaction I have ever known in life as well as all the improvidence, the extravagance and carelessness with regard to money, has come from the fact that I was, as Phoebe had already observed by the time I was seven, a little "teched."

I have made a great deal of money but I have saved none to speak of. In the years before and even during the Great Depression, I spent the money I made, more rapidly sometimes than it came in. I traveled where I wanted to go. I gave money away. I spent it on what turned out to be rash experiments. During the same period, I had friends who denied themselves the extravagances in which I indulged myself and my family, who stayed at home and saved their money and invested it, and when the Great Depression came, it was swept away and they had left neither money nor the good things which money could have brought them, in experience, in warmth, in richness, in memories. What I had acquired was not cash or stocks or bonds, all of which can be destroyed overnight by a depression or an inflation. Neither these friends nor myself had any money when the war and taxes and the Great Depression almost destroyed the fabric of our existence, but I did have left a great store of experience and memories and friendships in half the countries of the world out of which I could write and make a good living for the rest of my life even though I lived to be over a hundred. What I had was indestructible capital. I know now, in this tottering and impermanent world, how little folly my seeming folly really was. I take no credit for being motivated by wisdom, although the long experience of war and shortages and inflation outside my own country taught me many things which I was able to turn to the advantage of myself, my family and sometimes my friends. I acted as I did only because I was, as Phoebe said, "a little teched."

And so the whole adventure of the farm came too under the head of being "a little teched." A "smart" fellow would never have put so much money into so many acres of half-ruined hill land. A "smart" fellow would never have gone out and worked hard to earn that money. A "smart" fellow would never have attempted anything so extravagant as the Plan upon which Max and I embarked. And more than once the "smart" fellows have laughed at what we were trying to do. Fortunately I come of a hard-shelled family. Scottish and New England by blood, which was never very much concerned about the opinions of others. For generations, most of the members have set a course and to hell with the rest. Some have failed, in a worldly sense, some have succeeded, but I cannot think of one among them, either among the grandparents or the aunts and uncles or my own parents, who set out simply to "make money." I know too that whether the experiment succeeds or fails in the long run, whether I die in the big house I built or whether I die in a cottage with only an acre of ground around it, there will, at least be no bitterness and no envy, so long as I have a couple of dogs and a garden and that huge stock of adventures and memories. Among them the adventure in Pleasant Valley will be the brightest and most exciting of all. I also know that I will have left my mark on the surface of the earth in at least two places, one in the heart of France and one in Pleasant Valley in Ohio. There will be two spots at least which will be better and more productive and more beautiful because I stayed there for a time. I think all that comes of being a little "teched."

That is why for me, the memories of those two "teched" characters, Johnny Appleseed and Phoebe Wise, will always be respected. Both of them have become legends in our country because they represented something which all men and women at some time envy and seek to attain—that poverty, that simplicity, that richness which is the essence of true Christian teaching and far above all worldly riches. In a way the two of them acquired, like St. Francis, that intimacy with God and with Nature and birds and animals, that lack of all envy or ambition or greed, which is the ultimate "oneness with God." That is why they have become not only legends but in a way the saints of our countryside. They represented something which men yearn for, sometimes secretly, and with shame. One of them was a Swedenborgian, the other had no formalized faith at all beyond her love of all that had to do with Nature, yet they will be remembered in our county long after the millionaires, the Congressmen, the great manufacturers are forgotten. When factories are silent or in ruins and the industrial population is without food and perhaps shelter, the forest, the hills, the valleys, will still be there and the wild things which live in them will still be coming out of the forests and swamps at nightfall to harvest the wild, spicy apples, descended from the trees which Johnny Appleseed planted centuries before. About Johnny's legend and the hills and valleys there is an eternal quality, an ultimate refuge, which the mechanical civilization of man can never wholly approach or provide.

My father was always a little "teched." He was improvident and dreamy and never really succeeded at anything in terms of money and acquisitions, but until after middle age when he left the small town and country life for the city, he had a happy, rich and satisfactory life. In *The Farm,* I have painted of him a loving full-length portrait. His only significance here is the fact that he plays a great part in the establishment of Malabar Farm, for he fashioned and encouraged in me a love for all that had to do with animals and the out-of-doors.

My father undertook many things in life in order to feed his wife, his offspring and the visiting relations and to put a roof over their heads. He was at various times an oil operator, a bank cashier, an agent for the Great Northern Railroad in its campaign to open up the great fruit and grain country of the Northwest, the employee of a wholesale grocer's firm, an agent for the sale of farm properties. There was also a brief career in politics and for years he was secretary of the County Fair Association. None of these things made him rich; for his heart was really in none of them save the secretaryship of the Fair Association and that paid nothing whatever.

During all his life he was passionately interested in two things—the restoration of ruined farms, and the reformation of run-down or unruly horses. Of the latter we had an endless procession in my childhood and boyhood ranging from the pintos which he brought into the county from the West, to a team of wild and unruly Belgian mares which on one occasion ran away, destroyed a farm wagon, fifteen rods of a neighbor's fence and nearly killed my father and myself. Always there was something about the horses which needed reforming; they bucked or balked or had hard mouths or were vicious and unmanageable. They were the only horses I ever knew until I was nearly grown and my experience with them banished all fear of horses forever. Some of them we succeeded in reforming for both of us had the "teched" quality and understood animals, but a good many of them were hopeless. I suffered a broken arm and still possess a shoulder which can be dislocated at will as a result of attempts at reforming horses. Of course nobody made any money out of these reformation projects but we all had a lot of excitement and fun.

The restoration of run-down farms was no more profitable than the horse project and it usually tied up whatever liquid capital my father was able to raise. In my boyhood in our rich county there were farms which were already out of circulation through erosion or greedy farming. We always had two or three of these farms at a time and my father's efforts to restore them were primitive in comparison with what can be done in these times. At that time there were no county agents and few agricultural bulletins and the efforts and information of the Soil Conservation Service were primitive or nonexistent.

My father's ideas were sound as far as they went. He cleared away the underbrush that was already moving in upon deserted fields and reseeded with clover the fields which had been left bare by the last farmer to leave the place. And he fought gullies by filling them with cut underbrush and with rolls of rusted fence wire, and he spread the manure left in the deserted barns over the starved fields. Sometimes it had accumulated without having been hauled out for a period of two or three years and the beams above were worn smooth from the backs of the cattle rubbing against them.

Most of the restoration work was done by ourselves with the aid of small boys and sometimes girls who were friends of my sister, my brother and myself. There was never anything tiresome about the work; indeed it was never really work at all, for my father, by his spirit, and my mother with the good meals she prepared and carried along, invested the whole project with the air of a lark and a picnic. Every Saturday, Sunday and holiday and sometimes on days stolen during the week from the office where my father should have been working, a procession set out for one or the other of these ruined farms. It consisted of one and sometimes two buggies filled with axes, shovels, picks, food, women and children. Always there was a dog or two and a troop of boys on bicycles which careened up and down the road like a convoy guarding the buggies.

On arriving at the farm the horses were unhitched and turned loose to graze and we went to work, cutting undergrowth, building check dams, demolishing ruined fences and reassembling snake rail fences by picking out the good rails and burning the others. There was always the smell of wood smoke and of coffee and steak cooking under my mother's direction. The dogs ran rabbits and once or twice during the day the boys went swimming and in May, in the old abandoned woods and thickets, we found quantities of spongy morels which I discovered later in life were regarded by the French as the greatest of all delicacies. There was no need for any of us to be told that the "mushruins" were delicious, for in this world or the next there is no better food than morels cooked in butter and served with steak broiled over an open fire in the woods.

Evening came at last and, tired and sunburned and smoky, the whole procession started back to town, sometimes in the moonlight, those in the buggies singing old songs in chorus to the light of the moon while the tired smaller children slept soundly. Sometimes in the course of the day there came up one of the majestic and beautiful thunderstorms typical of that Ohio country and then everyone took refuge in the decaying barn to watch the spectacle of warm rain and lightning and black clouds swooping down quickly over the Valley and the wooded hills.

So the whole of my childhood was involved with ruined worn-out farms and even then, despite all the fun we had in the process of bringing them back to life, I was always aware of something tragic and awful about the manure-filled, half-ruined barns and the houses left by the last tenant who had not even taken the trouble to close the door when he left. Inside on the rotting floors beneath the half-ruined ceiling there was always an assortment of half-wrecked furniture and on the hooks along the wall, a row of worn-out Mother Hubbards and overalls. Mournful ghosts they seemed to me.

I did not understand then what all those farms and hundreds of others like them meant to the economy of the nation, nor did I speculate as to what became of that last tenant, moving to another dying farm or into the slums of an industrial city or worst of all, taking to the road with his family as an indigent tramp. I only knew that the desolation I saw was somehow evil and wicked.

SOME BIG IDEAS

LOOKING BACK ON THAT FIRST YEAR AT MALABAR IT IS ALMOST impossible to believe how much was accomplished. It could not have been done without vitality and faith and enthusiasm and youth; most of all, I think, youth. There were crops to be put in—corn and oats, wheat and soybeans, ramshackle buildings to be demolished, old ones to be repaired and remodeled, contours and strips to be laid out and terrace ditches to be built. Old fences lining the old square fields had to be torn down, fence rows cleared and new fences built to replace those along the lines of the new contours.

Youth played a large part because all the men who came there to work in the rehabilitation of the old farms were young. Max who ran things and taught me much, was thirty-two and he was the oldest. Pete and Kenneth and later on, Wayne, were under twenty-five. George was twenty and Jack only eighteen. Pete and Kenneth and Wayne were boys Max had known in his 4-H club and county agent work. They came of farming stock and already they had absorbed countless new and intelligent ideas through those excellent agencies the 4-H clubs and the Extension Service. George and Bill were neighbors. They came from over the hill, worked by the day and went home at night. All of them believed in what we were planning to do and none of them had any fixed, worn-out ideas to block our progress or their own.

And there was Hecker, the young Soil Conservation Service man who helped with advice and wanted the project to succeed, since he knew that nothing was so important to the conservation of the nation's natural resources as the scattered, sometimes, isolated "pilot" farms like ours where neighboring farmers had only to look over the fence and see the effects of treating the land properly. He knew, like others in the S.C.S. how the infection of good land use caught on in communities of intelligent farmers all over the country, how quietly, almost shyly, other farmers would begin presently to plow their fields properly and treat their wood lots with respect; how presently, as the infection spread, streams would begin to clear up, floods to cease, and fertility and profits to increase. He knew that farmers, especially farmers over middle age, are conservative fellows who must be shown and that many of them were skeptical of experiment station reports and government experiments, believing sometimes wrongly and all too often that because these things were accomplished through the expenditure of taxpayers' money, they were beyond the means of the average farmer and, therefore, useless to him. When they could look over the fence at a neighbor's farm and see him getting results, they would quietly and shyly imitate his methods.

One of the wisest of the countless wise plans undertaken by the Tennessee Valley Authority has been its system of "pilot" farms, by which the agency called for volunteers among the farmers of each county in the area to co-operate with it in working out a program of proper land use. Some of these "pilot" farms have been in existence now for seven or eight years and the results achieved have been beyond the wildest hopes or predictions. They have become islands of prosperity and abundance in whole regions which beyond the borders of the "pilot" farms were reduced to very nearly desert status. Each "pilot" farm has made its converts and slowly the whole landscape and aspect of those regions have begun to change.

Hecker knew all this.

Like so many devoted government *administrative* agents he was giving his intelligence, his training, his strength, his devotion to the job because he believed in it and preferred it to any other job on earth. Like many a county agent, he possessed an intelligence and training and character far beyond that of the men in other fields who made money many times his salary. Some day, I believe, the country will come to recognize the great debt it owes the nonpolitical men of the Extension Service and Soil Conservation Service and reward their self-sacrifice and devotion accordingly. Some day there will be fewer men sitting at desks in Washington and more in the field.

Besides Hecker there came to see us in the second year the able engineers and directors of the C.C.C. camp, arriving day after day with the boys to help in the building of fences. They were all colored boys and they worked like beavers and they built fences that will be there, staunch and strong, as long as posts and wire hold out.

And Ollie Diller from the Wooster Experiment Station came over to help us with the business of re-establishing the wood-lot areas on a sensible and paying basis.

The C.C.C. has been abolished, temporarily I hope, for of all the New Deal agencies it was, from the point of view of agriculture and natural resources, by far the most valuable. It never indulged, like the W.P.A. in the planless and senseless creek straightening that augmented floods and increased erosion or drained marshes that never should have been drained and had to be restored afterward. The C.C.C. had always a wise direction and a sound and efficient organization. We shall be paying for years to come for the damage done by the W.P.A. Why there should have been so vast a difference in the accomplishments of two more or less parallel government agencies, I do not know, but the evidence is spread across the whole of the nation. Perhaps it was simply a matter of direction, intelligence and ability.

There were others who helped, some of them government or state men, some of them practical farmers, some of them sportsmen, some simply intelligent men and women interested in natural resources who had good or wise ideas concerning fish or game or forests and among them many public-spirited citizens, perhaps the greatest asset of the nation, who will give money and what in these times is more precious, energy and time and interest, to any project which is aimed not to benefit a minority but the whole nation and its common heritage of democracy and good fortune.

Most of this aid and help, certainly that coming from government agencies, is available to any farmer. Too few of them take advantage of what is offered them by the Soil Conservation and Extension Services, by the land-grant agricultural colleges, by bureaus of animal husbandry. Some fail to seek advice and information and help simply through ignorance and indolence and not a few have developed a hos-

tility to all government agencies since the executive arm of government began to intrude in fields which should be administrative or legislative alone. Today any farmer, any farm boy can have an excellent agricultural education without ever seeing a college campus. He can learn an incredible amount through the 4-H clubs, the Future Farmers of America, the S.C.S., the county agent and other agencies and organizations. Nearly all the boys and young men who helped in the experiment at Malabar had taken advantage of some or all of these opportunities.

Around the barn of the Fleming place there had collected a whole cluster of outbuildings, most of them built in a planless ramshackle fashion by a succession of tenants as the farm had gone downhill. They were placed without much rhyme or reason, with no particular plan for saving labor, the only principle involved seemed to be that they were set far from each other in case there was ever a fire. The corncrib leaned lazily on its side. The roof of the hog barn had fallen in. The whole represented the story of a farm whose fertility had been mined out, whose tenants had never troubled to care either for the land or for the buildings. In it was the record, as well as the shiftlessness and the defeat of that generation or two of farmers who succeeded the pioneer days, when the richness of the soil had seemed inexhaustible, into the times which I recorded long ago in *The Farm* when it became necessary to adopt new agricultural methods and machinery, work harder, or starve to death on the land.

Throughout America there existed a generation or two of these farmers who tried to go along as their grandfathers had done, taking everything and giving nothing back, with the result that they had run both the soil and themselves into bankruptcy. In some places, in Iowa and Illinois and Indiana and northwestern Ohio where the depth and richness of the soil seemed inexhaustible, the old wasteful careless ways could still show profits although they were dwindling. That meant only that the farms could survive a generation or two longer, according to the newness of the land and the depth of the soil, according to the rate of theft from the soil and the speed of the inevitable disappearance of the topsoil.

West and north of us in Ohio and Indiana there are middle-aged and elderly farmers on the flat lands who tell you that all these newfangled notions are so much rubbish. They say and believe that they were losing neither fertility nor topsoil. Yet there are two terrible and concrete evidences. These same farmers, who were forced to tile their flat land for draining purposes, are today plowing up the tiles they planted only twenty-five or thirty years earlier. And there is the terrible evidence of Sandusky Bay.

In northwestern Ohio the land is flat and one can see for miles across the country without a perceptible roll in the land. It is the newest land in Ohio and some of the most fertile in the world. Once the whole area had been lake bottom, a part of Lake Erie, and then for a few centuries it had been a vast marsh inhabited by bear and wolves and wild duck and geese.

A little before the Civil War the whole area was drained by a system of vast ditches and the land sold off at a dollar an acre to settlers. It is cornland and the best baby beef in America comes from it. Thousands of steers are "fed out" there every year and the fertility has been kept high by the millions of tons of manure that goes back into the soil every year. The yields of crops, particularly corn, are prodigious. Almost any farmer in the region will tell you that his soil is ten or twelve feet deep and that his land is so flat that there is no erosion. Yet the time is not too far off when all that region may once again become marshland, for the Black Swamp area is slowly returning to its old condition as the soil wears down to the level of the drainage ditches and the big lake to the north. You cannot see it go; evidence of the change might never occur to a farmer until he begins to plow up his drainage tiles and even then he will try to persuade you and himself that this is not because the soil is wearing off but because the frost or the "working of the soil" has brought the drainage tiles to the surface. But the terrible evidence is Sandusky Bay.

The whole area lies in the watersheds of the Maumee and Sandusky rivers. The Sandusky flows into a marsh-bordered bay about thirty or forty miles long and from four to five miles across. Within the memory of a man forty years old, some of the best fishing in the world existed in that bay—bass and pickerel and all sorts of native game fish. Today there are in it only the sluggard mud-loving carp and a few perch because the water of the bay is seldom clear any longer. The soil of the drained great swamp has been moving into Sandusky Bay since the Black Swamp was drained and put to the plow. The bottom now is mud and the areas along the edge grow shallower each year as more and more silt is deposited. Meanwhile the game fish have left the bay. Once it was an important spawning area for the lake fish which provide Ohio with one of its important industries. The fish no longer spawn in the mud-filled bay and the effects are being felt in the fishing industry.

One day the farms of the rich Black Swamp country will have moved into Sandusky Bay, filling it up and leveling off the land of the rich watershed behind it. One day both bay and farm land will be level and the Black Swamp will return.

It is true that in this area there are good farmers who practice proper crop rotation and cover their bare cornfields in winter with root blankets of wheat or rye and they are doing a good job of anchoring their soil. Very little of it is going down into Sandusky Bay. But there are

others who leave their fields bare to wind and frost and rain. They will tell you that you are crazy, that they have no erosion on their flat land. When a farmer in the South, the East or the Middle West, tells you that he has no erosion by wind or water on his bare fields and that it is unnecessary to take any precautions, you may put him down either as ignorant or a fool. The truth, which any man can see, is that a bare field is abhorrent to Nature and she sets about at once to blanket it with vegetation or to destroy it. Winter-bare corn, cotton and tobacco fields cost this country annually millions of dollars in the loss of soil.

On Malabar, in genuine hill country, where ugly gullies and denuded hilltops tell their own story, the evidence was all about us, yet there were farmers of the last generation, unlearning and unwilling to learn, who thought and said the measures we were taking that first year were crazy. It is true that they were not very good farmers and that in the case of most of them, they will probably be the last occupants of the land they were farming. Bankruptcy and the forest will move in again.

Behind that philosophy lies a large segment of the history of the United States, once a vast wilderness of incredible richness inhabited by a few hundred thousand half-savage redskins. White men from Europe came to it and set out to pilfer rather than to develop it. The riches seemed inexhaustible and a tradition of farming grew up among the frontier men which made of the American farmer one of the worst farmers in the world. The tradition and habit was simply that of mining the land.

The formula was simple. First you simply cut off or burned over the forest or prairie and then you went to work wresting the fertility from the soil in terms of crops as rapidly as possible. Sometimes the fertility or the topsoil lasted two or three generations, sometimes longer. Then when the soil was worn out you went west to Ohio or Indiana and repeated the formula. When land was exhausted there, Iowa and Kansas and Dakota lay ahead, and finally Oregon and Washington and California. The good land that could be had for little or no investment and could be "mined" seemed without limits. "The West" became a byword for opportunity—opportunity principally for more free, rich, virgin land. Very often men went west to take up land less good than the land they had recklessly destroyed. Sometimes they exchanged a mild climate for a harsher one, well-watered areas for country afflicted by drought, or waterless land which had to be irrigated. They could have done better and been happier and more prosperous and comfortable if they had cherished the good land they destroyed and remained on it.

Like a plague of locusts they moved across the continent, leaving behind here and there men who found the soil so deep and the mining so inexhaustible that there was no necessity for migration. And here and there they left behind a good farmer, wiser than the rest, who cherished his soil and farmed well. But the good farmer was and is as a rule a "foreigner"—an American whose tradition and training in reality went far back into a Europe where there had been little or no cheap land for five centuries. By them their piece of land was regarded properly as their capital; and an intelligent or a wise man does not throw his capital out of the window. As a rule the more recently arrived the immigrants, the more they respected the piece of land they were able to acquire in the New World. Very often they acquired farms ruined by farmers of old American stock and restored them. Too many of the descendants of the older stock followed the wide-open reckless traditions of the American farmer in which land was not a capital and treasure but merely a speculation or a "mine." The principal exceptions among the old stock were the Pennsylvania "Dutch," the Amish and the Mennonites who lived closely among themselves, holding fast to their particular variety of religion, to their customs, even to their language. They stayed on the land they settled upon and made it richer and more valuable each year by farming well.

Not all of the fault lies with the farmer himself. As Americans, we are all immigrants to a new world and since most of our early stock came from central or northern Europe, the traditional agricultural methods of that region were brought here with them. The earlier stock not only came to a country of apparently inexhaustible resources but the agricultural methods it brought with it were hopelessly unsuited to the different climate and soils of America. The inexhaustible richness made our farmers of old American stock reckless and greedy exploiters; largely speaking, the older the American stock the poorer the farmer. Those immigrants who arrived more recently have managed to preserve, like the earlier religious sects, their reverence and respect for the soil.

The use of European agricultural methods in the American climate has been disastrous. In the temperate areas of northern and central Europe there rarely occur the cloudbursts, the thunderstorms, the violent winds, the seasonal droughts and floods, the temperatures ranging from twenty below zero in winter to one hundred degrees Fahrenheit in summer which are commonplace in the American climate. In the European area the climate is more temperate and the rains fall gently, sometimes as in Normandy and in the Channel Islands, in a form of more or less steady drizzle. In all the eighteen years of my experience with agriculture in France, I saw only once a thunderstorm and cloudburst as violent as the kind of storm that happens a score of times every summer in our Ohio country.

As the wilderness was subdued, during the progress of the white man westward, forests were cut down, high prairie grass plowed under and millions of acres of the best mixed legume grazing land in the world was plowed up, overgrazed and burned over. Swamps were drained and streams straightened, sometimes senselessly. Within a period of from fifty to one hundred years the whole of the vast Mississippi Basin was changed almost beyond recognition by the hand of man. Much of the land once covered with forest, grass and marshland was left bare to cloudbursts, thawing snow, tornadoes and other violent manifestations of a climate much more like that of China than of Europe. The result was devastating floods, dust storms, droughts and lowered productions in some areas approaching desert status.

Dr. Hugh Bennett of the Soil Conservation Service estimates that if the soil lost annually by erosion in the United States was placed in ordinary railway gondola cars, it would fill a train reaching four times around the earth at the equator. At the Georgia State Agricultural College tests made upon one acre of ground farmed by the conventional method used in cotton cultivation in that state showed an average loss of 127 tons of topsoil a year over a period of five years.

About twenty-five years ago there began a spontaneous movement toward finding a system of agriculture more suited to the American climates and soils than that imported from Europe. It came none too soon. In this search countless people took part, most of them working individually, experimenting, watching the earth, working always toward a common end. There were market gardeners, garden club members, government bureau men, agricultural college professors, city farmers, dirt farmers, schoolteachers, all working individually toward the common end of finding an *American agriculture*. Only in the past few years has it become apparent that an actual revolution in agriculture had been in progress for a long time. I think that the publication of *Plowman's Folly* by Edward Faulkner did much to make the character of the revolution evident. It is certainly true that never before in the history of the nation has there been so intense and so widespread an interest in agriculture.

Slowly but certainly a system of agriculture suitable to the United States has been evolved. It has grown out of the remote past of Asia and the Near East, out of discoveries made in half the nations of the world, out of experiments and the brains of countless devoted and intelligent workers. The system includes terracing and cover crops, trash farming, proper drainage and forestry practices and pasture treatment, the use of legumes as green fertilizer, diversified farming and the rotation of crops. The revolution is still in progress, growing and expanding. It is not only important to the people of this nation, but to the people of other nations with similar problems of soil and climate. South Africa, Palestine, China, Mexico, even so new a country as Venezuela, have called for and received the aid and experience of Drs. Bennett and Lowdermilk and their staffs from our Soil Conservation Service. Perhaps of all the aid given so lavishly to foreign nations by our government, none will prove so valuable as that given, without fanfare, with out special reward, by the men of our Soil Conservation Service.

To many a reader it may seem fantastic to talk of shortages of food in the United States. No doubt it seemed foolish to talk of such things in India or China a few centuries ago, but today half the populations of India and China are born and die without ever having had enough to eat one day in their lives. If today you shut off countries like Denmark or Italy or England or Germany from outside markets, the diet of the peoples of these countries descends to the level of malnutrition and even starvation. In all those nations there is not enough good land to feed their populations. In some of them the good land never existed. In others, as in India and China and southern Italy and North Africa, the good land was destroyed by the natural processes of erosion, by bad agriculture and by the dislocation of populations which follows wars and revolutionary changes.

Nearly all of these forces are at work in this country, some violently, some in an imperceptible fashion. Despite all assertions to the contrary, despite all the talk of abundance, despite extraordinary production records of the farmer, we have not been able during the war to provide all the food necessary to keep the diet both of servicemen and civilians on the level to which this rich country was accustomed in peacetime.

There have been, from time to time, shortages of milk, of meats, of eggs, of vegetables and fruits, all of those high-protein foods which provide the energy and create the physique for which Americans, as a race, have been famous Those high-protein diets which can alter the very physique and character of second and third generations descended from undersized stock in Europe. A nation's energy, intelligence, and initiative is no greater than the citizens who go to make it up, and the individual is no better than the food he consumes and the soil on which he lives. The poor physique and intelligence and initiative of the southern Italian, the North African, the poor white and the sharecropper of our Southern states is born of poor diet and worn-out soil, bereft of the phosphorus, the calcium, the minerals which go to make better physical strength and intelligence and good citizenship. As our soil grows depleted, as our diet becomes more limited, the stamina of a nation or an individual goes downhill. The evidence that poor soil makes poor people and poor people make poor soil worse and so on *ad infinitum* is written across the face of the whole of the earth and of mankind itself.

In our own country there is no more virgin, well-watered land available. Land that can be irrigated or drained is not always good agricultural land, and often enough in our West, the very water used to irrigate the land is so alkaline that it ruins the soil within a few years. Fully a quarter of our good agricultural land has been already reduced to the lowest status, fit only for reforestation and sometimes not even fit for such a purpose. Fully another quarter is in an intermediate stage of destruction. There has been much talk of settling returning veterans on the land, but what land? The good land is not for sale or if for sale it is expensive land. There are no great prairies with deep, black soil waiting for settlement to absorb returning soldiers as it absorbed them after the Civil War.

Meanwhile our population continues to increase while the productive capacity of our land decreases rapidly. It is probable that not 5 per cent of our agricultural land produces anywhere near the 100 per cent of its potentiality. It is also probable that about 70 per cent of our agricultural land produces not 30 per cent of its potentiality. The farmer's problem is not entirely one beyond his control. Much of the rural poverty and insecurity arises from the fact that there are too many bad and careless farmers and too many lazy ones who are content to live as their pioneer grandfathers and great-grandfathers lived, working during the crop season on single cash crops and sitting by the stove or in the village store all through the winter. Their forefathers had deep virgin soil which could support that sort of existence and when the land wore out they could go elsewhere. Today in order to succeed, no matter what the prices or the parity support, a farmer has to be both intelligent and informed and to work and work intelligently with good modern farm machinery. This is no longer a new country with limitless resources. Each day conditions come nearer and nearer to approximate those of Europe with respect to agriculture, population and economic security. This need not be so; it is so largely because we have been wasteful and reckless in the treatment of our natural resources.

It might be said that if every farm produced 100 per cent of potentiality, there would only be in certain areas a glut of food and lower prices for the farmer. Under existing conditions that is probably true but there are two elements at least which, if corrected, could serve as checks.

One is proper distribution both in volume and extent as well as in reduced costs. The other is the buying power of the nonagricultural population. As to the first, it is abominably badly managed in the United States as indeed it is throughout the world. Costs of distribution are overhigh, handicapping the farmer who comes out nearly always at the small end of the deal. And distribution is determined largely in a haphazard fashion, seeking always the high-priced market, until it becomes glutted and the price collapses. The fact remains that about 50 per cent of the population of the United States suffers from malnutrition or at least undernourishment even when there is plenty of food. Part of this is true because of ignorance and bad diet and part because of bad distribution and inability of the people to buy what they need and want.

The extreme example of imbecility of bad distribution occurred during the Great Depression when in cities throughout the whole breadbasket area of the Middle West, unemployed were fainting in the streets from hunger while a few miles away farmers were burning wheat and corn for fuel and killing off their pigs in order to raise artificially the prices of food for people who already could not afford to buy it in sufficient quantities. Like so many New Deal measures the whole agricultural restoration program was largely improvised and superficial. It never struck at the root of the matter until Milo Perkins conceived the idea of food stamps as a form of relief and of getting excess food supplies from the producer to the people who needed food and had not the money to pay high prices for it. Before this reasonable and intelligent plan had a chance fully to prove itself, the war boom solved the problem of the unemployed and high wages removed the fundamental necessity for it.

The same imbecility of inadequate distribution holds true as well in an international sense. It does not argue well for man's intelligence that the populations of Russia, of India or of China should starve while American farmers are receiving for what they produce less than it costs to raise it. There has never been any intelligent effort to distribute food internationally as it should and could be distributed, so that every nation would have out of the world's combined agricultural production enough to eat. Possibly if the food supply produced by American farmers were properly distributed among the American population in terms of proper diet there would not be too much to eat per capita, but if there were any surpluses there is plenty of place for it in a world where millions of people never have enough to eat. Russia, China, India, most countries in the world have commodities for which we have a growing need and which we actually import. To date the emphasis has all been on the exporting, not of food from this country but of manufactured articles protected here at home by high tariffs. Nobody has paid much attention to the problems of proper world distribution of food as a means of sustaining the farmer's prices and absorbing his surplus. On the contrary during the past twenty-five years we have imported ten per cent more food than we have exported.

We have come, it would seem, a long way from the run-out soil and the ramshackle buildings of the Anson place, but in reality not so far as it would seem for in those worn-out fields, in a farm which had ceased to be a national asset and had become a national liability, lay the answer to many of these things. Around the stove in the miller's house we used to talk about them, night after night between games of euchre or hassenpheffer.

We knew that to farm the Anson place we would be forced in the beginning to employ time, labor and machinery and fuel to produce on two hundred acres what could be produced with a quarter of the time, labor and fuel expenditure on fifty acres or less of good, fertile well-managed land. We should have to continue this waste of time, labor and fuel until that two hundred acres of land was restored to its maximum fertility. Our problem was exactly that of the great majority of American farmers—the problem of wasted energy and labor costs. Seated by the big stove I used to tell of agriculture in France where the aim of every farmer was to make each acre produce the maximum of its potentiality without loss of fertility. That, it seemed to me, was the best formula for good farming I had ever encountered, yet to most American farmers it is a strange one. As Jack said, "I never thought of it that way." Nevertheless, it is the way we shall have to think of it one day if our agricultural economy is to stand on its own feet without subsidies that drain the profits from the rest of our economic life.

As our resources in metals and mineral oils dwindle, it is becoming more and more necessary to grow our resources in terms of trees and of plants from which plastics, oils and fabrics can be manufactured. The topsoil and the general fertility of our land are being dissipated steadily while the demands made upon it both by food and by industry are constantly increasing. Without soil, good soil, well-managed soil, these demands cannot be met and the squeeze will make itself felt more and more in our living standards and national economy. It is my own belief that the squeeze is already making itself felt.

It is not good enough to argue that we are already producing too much cotton or too much wheat. The fact is that we should be producing less cotton if there is not enough demand for it, and that we should be producing all the cotton we need and for which we can receive a decent price on approximately one-tenth the land now employed in producing this commodity. That could be done if proper land use was practiced. The soil released from cotton culture could be put to raising other crops for which there was greater demand—for cattle, for tung oil, for hemp or any one of the dozens of products needed for industrial purposes, or simply be added to the acres producing forests which are destined to play a greater and greater part in our industrial life and economy.

These were big questions for a bunch of farm boys to be discussing around a stove in Ohio, but it is encouraging that such things are being discussed more and more frequently and intelligently by the younger generation of farmers. Much of the farmer's independence and individualism is born of the old pioneer tradition of the frontier, where he *had* to be self-sufficient and was by circumstances isolated. It made him

shy and gave him a feeling of inferiority in the realm of affairs which did not touch him immediately. These feelings have too largely colored the American farmer's point of view in the past. They gave rise to the distinction between "dirt farmers" who were pictured as unshaven men in overalls and straw hats, chewing tobacco and a straw at the same time, and "city farmers" who drove about in limousines and kept registered herds in plush stalls. These distinctions are, I think, beginning to vanish because of the radio and the automobile, because most prosperous farmers today have plumbing and electricity. They are interested in national and world politics and discuss them. On the other hand, the old "city farmer," like the traditional hick, is beginning to disappear because he can no longer afford an "estate" but more, perhaps, because of his growing genuine interest in sound agriculture and stockbreeding. The old lines are beginning to vanish and the old frontier philosophies and traditions are beginning to break up. This is all to the good and means that in agriculture and as a nation we are achieving maturity.

I am aware, of course, that what is written above concerns largely the farmers of the rich agricultural areas of the nation, and that in the agriculturally wrecked South and in similarly ruined or naturally poor areas over the whole of the country, there still exist millions of miserably poor farmers, living at or below a subsistence level This more than any other fact of our agricultural life and economy is directly related to soil—either naturally poor soils or soils wrecked by man's ignorance and greed.

Sentimentalists and reformers have made much of the wretched condition of the tenant farmer, the sharecropper, the farm laborer of the Southern and border states, and rarely have they exaggerated the situation. The measures for rehabilitating these people have been almost entirely upon educational or economic lines, both of which solutions are in the long run superficial ones. In almost every case, these unfortunate people are not simply victims of absentee ownership, of an antiquated and vicious economic system and poor educational facilities. They are, primarily, victims of malnutrition, bad diet, and above all—poor soil! No matter what the heritage of blood you cannot make an enterprising and successful farm population of people who not only exist upon a wretched and unbalanced diet but upon a diet of food grown on miserably worn-out soil. You cannot make an intelligent, healthy, energetic and prosperous population, nor a population even capable of education, of people who haven't enough calcium or phosphorus in their bodies. You cannot get calcium, phosphorus and all the other elements and trace minerals out of poor or worn-out soils where they no longer exist.

Most of the rural population of the Southern and border states live upon worn-out soils from which the minerals vital to human health, energy and intelligence, have disappeared either through a vicious and greedy system of agriculture or because the soil was poor and thin in the beginning and its richness was soon exhausted. No amount of subsidies, no change in economic conditions will ever alter the tragic and pitiful condition of the people in these areas so long as bad diet and miserably poor and worn-out soils are the physiological bases of their existence.

There is in our county not far from us a whole family of five generations standing, which in itself provides a case history of the degenerating effect of bad diet and poor soil. Let us call them Smith. The first

Smith came from Germany and was a good farmer and settled on virgin land in the hills. He married an American wife and their sons were less good farmers than the father. In the succeeding generation, the land, not only of the original farm but of the farms worked by the sons, grew poorer with each generation through erosion and the process of leaching and robbing the soil. With each generation, as the soil grew poorer, the income diminished and the diet became a little more meager, until presently it consisted largely of meat and potatoes and not much else. But the few vegetables and fruit eaten by the children had less and less calcium in them, less and less phosphorus, less and less magnesium and other minerals.

As the children grew up they lacked the intelligence, the energy and the health of the first and second generations. Consequently, they grew more listless and less effective as farmers and their soil grew still poorer. The endless vicious circle has left in the fifth generation, nine male descendants of the original stout, healthy immigrant. Of these, two still cling to worn-out farms and their children are subnormal as to health and intelligence. The likelihood of their ever being more than liabilities to the community or the nation is extremely slight. The other seven work as day laborers or migratory workers, having long since abandoned their worn-out farms, but their children are perhaps better off than the children of the two who still cling to their wretched farms, for their diet is a little more varied and some of it, bought in town or village stores, is certain to come off soil which can feed their small bones and brains with the minerals which are essential to health and intelligence. The striking fact is that in the same Valley exist today, families of stock no better than that of the original Smiths, who through good farming and good diet and good soil have kept their vitality, their strength and their value to the nation as citizens.

From time to time during the life at Malabar, one or the other of the fifth generation Smiths has worked for us. Their characteristics were pretty much the same. Any one of the Malabar boys was worth at least three Smiths when it came to work, to initiative, to intelligence. The sad thing about the Smiths was that they could not help themselves. There was no will to work because in their bones and bodies and glands there was not enough of those minerals which made for human health and intelligence and energy. I think they would have preferred to be prosperous, hard-working members of the community but their minds were dull, their bodies perpetually tired. One of them, Henry, could have made a remarkable naturalist for he knew everything there was to know about the wild birds and animals and plants and trees of the Valley. He knew things which I am certain some of our most distinguished botanists and ecologists do not know, but with Henry, there was neither the education nor the energy that would have permitted him to organize his knowledge and make it of use to science. Poor Henry was simply the victim of poor diet and worn-out soil.

Later on, during the war, when the shortage of farm labor became a problem on the Ohio farms, the government organized a migration of workers into Ohio from the border states hill country of Kentucky, West Virginia and Tennessee. Partly through my own insistence the men and their families were handpicked as the best available stock by their local county agents and A.A.A. men. They were given a six weeks' special course in the mysteries of farming as it is practiced in the rich Ohio country, but the experiment was almost a total failure. Some of the trouble arose from a difference in manners and social habits but most of it came from the dissatisfaction of the Ohio farmers with the incapacity of the emigrant workers for work and responsibility. Eventually the experiment was abandoned altogether.

There was no essential difference between the stock of the border states hill country people and that of the Ohio farmers. One group had been nourished on its traditional bad diet and upon miserably poor and deficient soils, the other on the rich varied soils of Ohio. An earlier attempt at bringing Southern hill country people into Ohio had been made during the depression when farmers sought cheap labor, and the results had been the same. Some of these people came to work for us periodically and contact with them made us understand more vividly than any other factor the problem of the poor farmers in the South, and convinced us that the problem was as much one of diet and soil as of economics or social systems. They were wretched physical specimens, fundamentally unsound, with poor teeth or no teeth at all, both lazy and irresponsible and in one case, at least, without a vestige of conventional morality.

In the case of Lester and his wife we made a valiant struggle to do something for them. The bones, the brains, the whole physique of the parents was deficient but already established beyond much hope of change. Lester, at forty-four, had exactly one tooth in his head and the only purpose it served was to provide periodically an excuse to visit the dentist, a course we urged upon him. I don't believe Lester ever saw a dentist. Each time he returned from the hypothetical visit, he was drunk. There was no need nor any purpose in attempting to discipline him. His wife took care of that with a thorough cursing and beating. Never had a woman a finer flow of language or, despite bad soil and diet, a stronger arm. But the five children, all under seven, were worth saving. From them one could see what the parents must have been, what potentialities they had once possessed; potentialities ruined and destroyed simply by poor soil and bad diet. All five of the children were good-looking and bright and quick. They provided ample evidence that the blood lines were sound and good as is so often the case in the Southern agricultural areas. For most of their lives they had been living on good Ohio soil as the children of migratory workers and so they had health, despite a bad diet, which undoubtedly was superior to that of their parents at the same age. Certainly during their brief stay with us they had the best of eggs and milk, as well as vegetables and fruits, all grown on good soil to which minerals and humus had been at least partly restored.

Our adventure with Lester came to an end when his depredations upon our tools, our eggs and other property extended to those of neighboring farms and finally the sheriff took a hand. Lester was let off without punishment on the guarantee that he would leave the county, a common method of dealing with border mountain people who found their way into Ohio. When last heard of Lester and his family were heading south. He did return long enough to vent his resentment upon our prosperous community by breaking all the windows in the house we had rented and repaired to shelter him and his family.

The sentence of banishment was, of course, no solution. It rid the county of Lester but only wished him onto some other community. There was nothing much to do about Lester and his pugilistic wife. They were the victims of physical malnutrition and bad economic and social environment; but the children were potentially good and valuable citizens, any one of whom, with proper diet and nutrition, might one day make a contribution to the good of the nation. I am afraid that eventually, as Lester is shunted from one county to another, he and his family will end up once more on the poor worn-out soils of Arkansas whence he came, and the five, bright, good-looking children will end up as victims of malnutrition and poverty—the poor people who make poor soil worse which in turn makes still poorer people.

Some of the hill country migrants have stayed in our county. In the beginning they lived in shacks in isolated communities, more or less pariahs among the prosperous farmers. Gradually some of them living on new soil with novelties such as milk and green vegetables and fruits grown on good soil in their diets have changed their characters and acquired small holdings of land which could produce not only health and vigor but a decent income, many times that of the average $168.00 a year cash income of the agriculturally wrecked regions from which they came. There is an absorbing task for some government agency to record the changes in the physique and character and even morals of good American stock transplanted from miserably poor soil to good soil.

It is probable that a farm worker in the Middle Western breadbasket country is worth four or five workers from the poor soils of the Southern states on the basis of energy and working capacity alone. On the score of responsibility, initiative and intelligence, the ratio undoubtedly runs even higher in favor of the Middle West. The breadbasket area is not, however, without its victims of worn-out soils as the case history of the Smith family has shown. As the fertility of soil declines so does the vigor and the strength of the people living upon it. The record is there for everyone to see in every agriculturally worn-out area in the world—in large parts of India, in North Africa and Mediterranean Europe, in parts of China, in the South of the United States, and even in as newly settled a country as Venezuela.

Our own history of bad agriculture and worn-out soils is almost a record among the peoples of the world. Once this country was in the habit of boasting that it had among its citizens no peasants but only farmers. Today that boast is no longer true. Today a large proportion of our small farmers exist upon an income and living standards below those of the average European peasant and in some regions we have farmers whose income and living conditions are scarcely better than those of a Chinese peasant. They exist upon lands which should have been left in managed forests, lands which were never suited for agriculture and were worn out and destroyed within a generation or two, or they exist upon rich soils which were destroyed by greed and bad farming within a century or less. The problem is not only an economic one; it is a social one as well for the rural population of the nation is not increasing from the stock of good, healthy farmers established upon good lands but from the wretched "peasant" population settled upon poor and worn-out lands. In the rich Middle West, farm families have rarely more than two or three children; in the poorer regions families are likely to increase at the rate of one child a year or more. This is something to think about.

On the hill across the Valley from us there stands a big fourteen-room brick house and near it what is left of a big barn. The big house, built of pink brick burned on the place, stands empty, the broken windows staring out on ragged deserted fields grown up with goldenrod, poverty grass and sumac. Last year the big barn, tired and deserted for too long, collapsed. In the Valley it is known as the Mason place.

The place is owned by a man who lives in a distant city. He keeps it for sentimental reasons because it is the cradle of his family in the rich Middle Western country. Last year he revisited it and after seeing what we had done to restore farms like it, he offered us the farm, rent free, in the hope that we should by good practices restore some of its value. Today it has none. It produces nothing. No one wants to buy it. If it were offered at sheriff's sale no one would make an offer. We went over it carefully and came to the conclusion that it was not even worth fencing in order to pasture sheep. That farm is finished, fit only for reforesting in pine trees. A few hardwoods—beach, maple and oak have seeded themselves in the gullied fields, but they do not flourish. There is too little soil.

The place haunts me. Occasionally in the night I wake up and think about it—the big, beautiful brick house, dead and half ruined and the vast barn, a collapsed heap of firewood. It is like too many farms in America.

Slowly I have been unearthing its past. In the Valley it is not too difficult to do because there are still old people who remember a long way back and have not forgotten the stories told them by.their fathers and grandfathers.

There is nothing very unusual about the history of the Mason place. It is the history of millions of other farms in America.

The house was built in 1820, about fifteen years after Ezra Mason his wife and ten children came to settle in the Valley. It was forest country and that pioneer family cut off the virgin trees, piled them in great heaps and burned them. When the trees were cleared away there remained about ten inches of rich black forest loam, the residue of a million years of growth and decay. It grew wonderful crops and in a few years Ezra Mason and his sons burned the bricks which built a handsome house and a big barn to shelter the cattle, sheep, horses and hogs during the winter months.

Ezra died at last and he left the farm to his oldest son and divided his money among his other children. There was enough for each of them to go further west and establish themselves on rich new farms of their own. The wealth came out of that thick, rich black loam and served to develop more American resources. It helped to build the nation.

Ezra's son carried on and built additions to the big brick house and the barn. There was a little less soil. It had been farmed greedily and much of it left bare to the elements had washed away, but there was still enough for Ezra's son to follow his father's example. He left the farm to his eldest son and divided his money. His children went west to Iowa and Wisconsin, to Kansas and Missouri.

With the third generation there was less soil and less money. Two or three of the children went to the city. In the fourth generation there was still less soil and less money. And then the Masons moved away altogether and left the place to tenants who farmed it greedily and carelessly, and at last there came a day when no tenant wanted it and the fields were rented out to neighbors who took everything off the land they did not own and put nothing back. The good loam topsoil was almost gone. It remained only at the bottom of the fields or in the depressions. Everywhere there were gullies. And then came the final stage, when the land was no longer worth plowing. Tramps and squatters lived in the house until the roof decayed and then they too moved away, and the Mason place died.

A farm like the Mason place makes you do a lot of thinking. At one time that land supported as many as twenty people, providing them with good food, with clothing. From it well-nourished, sturdy children went to school and to the Valley church. It produced and sold eggs, butter, milk, beef, pork, sheep, wool, chickens and grains of all sorts. It deposited money in the banks and borrowed money on which it paid interest. It bought books, clothing, farm machinery, lightning rods, harness, carriages, organs and any number of things. Each year the money it spent in the neighboring town found its way into banks and circulated over the whole of the nation. From among the well-nourished children there came a bank president, a Senator, a governor and any number of schoolteachers and lawyers and general good citizens. All of that came out of the rich black loam it had taken Nature a million years to create. Once the Mason place was a rich economic asset to the nation.

Today it is a tragic liability. It buys nothing. It produces nothing. Its worn-out deserted fields contribute their share of runoff water to floods which every year cost the nation millions of dollars' damage. It is only by chance and sentiment that it pays any taxes, and even the taxes are not paid by that land but by a factory in town hundreds of miles away.

When you think about the Mason place you can't help thinking that there are three or four million other farms in the same condition scattered over the whole of the United States, farms which are no longer economic and social assets but grave liabilities. Another million or two are on the way to becoming liabilities.

Before the war there was a population of migratory workers totaling about eight million men, women and children. Most of them came off farms like the Mason place. They worked on an average of three or four months a year and were on relief paid for by the rest of the population, for the rest of the year. They paid no taxes. They were a drag upon the national economy. Their children had little or no schooling and little family life. They migrated from place to place, living in shacks and jalopies. Their children were virtually being trained to become vagrants. A few years ago they descended upon California in such numbers that disorder and near civil war was the result.

Temporarily, much of this population has been absorbed by industry and the men by the draft, but immediately the war is over they will be back with us again, because their problem is not that merely of booms and depressions. It is a result of the permanent illness of the land and of American agriculture itself. Below the level of the dispossessed migratory worker is another whole population, perhaps even worse off—the tenants and share croppers still anchored to worn-out agricultural land who have not the money or energy to board a jalopy and take to the road. They pay little or no taxes. They produce little more than they themselves consume. They buy only the cheapest and barest of necessities. Very often their schooling and local government are paid for by taxes contributed by more prosperous elements in their states.

A good deal more than half our population lives on farms and in villages and small towns dependent upon agriculture for their economic existence. While there are ups and downs in agricultural income, the general trend is downward, because our land steadily grows poorer or is being destroyed altogether at a shocking rate. The living standards of a nation are based primarily upon its agriculture, just as certainly as is its food supply. More than half of our population in villages, towns and farms, buy automobiles, breakfast food, radios and all the long list of commodities and of income directly from the land or derived from it through trading and servicing. When their income sinks, buying declines and eventually ceases, with the result that the production of factories declines and unemployment gains.

Not only is the land and agriculture the source of much of our wealth but the very base of our economy. Each year more and more county and village banks go out of existence because the agricultural land in the vicinity has become so poor that it can no longer deposit money or stand as security for interest-paying loans. In each depression in our history the disaster has begun at the agricultural base and eventually brought down the whole of the economic structure. Progressively in recent years this situation has become worse. It will continue to do so unless our program of soil conservation, restoration and the distribution of agricultural products is not more widely understood and expanded.

For the past twenty-five years despite installment-plan buying, volumes of statistics and advertising ballyhoo, our boasted high standard of living has been slipping downward toward the level of that of Continental Europe. Relief, vastly increased taxes, higher prices, food scarcities, depressions, are caused by a variety of reasons, but among them all the most serious is the sickness of our agriculture upon which is based all the rest.

Across the Valley stands the gaunt, empty, beautiful brick house of the Mason place, and each morning when I look across the Valley and see it, I think of all the things I have written above. The Mason place stands there as a symbol of our wasted and dying land, of our sick agriculture. There is not one citizen in the whole of the United States who is not affected by the Mason place, by the disaster it symbolizes in taxes, in higher living costs, in economic depression, in water supply, in floods and in thousands of ways because now, as since the beginning of time, in this nation as in every other, soil and agriculture are the foundation of everything.

Migrations of farm people from ruined farms to industrial cities is no solution of the general over-all problem; it only increases the seriousness by adding to the great industrial population in crowded cities which is without economic security and in time of depression becomes overnight a frightening burden upon the economy of the whole nation.

Recently there have been strong evidences of a spontaneous movement toward the decentralization of our vast and ugly industrial cities. During the mad scramble of industrial development during the nineteenth century and the beginning of this one, these cities grew up without planning or foresight, simply because here there was a harbor or there a point where iron ore and coal and limestone were available at cheap rates of transportation or because railroads and shipping provided convenient shipping facilities. Into them were poured the streams of cheap labor drawn from among the politically and economically oppressed classes of Europe. In these great cities the immigrants lived together, segregated according to race or nationality—the Jews, the Poles, the Yugoslavs, the Italians and so on, speaking their own languages, reading their native-language newspapers, retaining all too often in the first generation or two a closer allegiance to their native countries and the political and social ideas born there than to their adopted country and the ideals of democracy. Very often they were sheltered in tenements and shacks in surroundings even more sordid than those they had left behind in Europe. Children grew up in communities where, save for the church on Sundays, the only social life centered in saloons, poolrooms and brothels. They lived on street corners and attended and still attend overcrowded, unsanitary and insufficient schools. It was inevitable that there should grow up out of these surroundings whole groups of gangsters and racketeers as well as radical and anarchic social and economic ideas which had little to do with the democracy of the new nation to which the immigrants had come in search of liberty, of better economic conditions, of hope for a better place in life for their children.

The responsibility lies much less with the individuals who comprised the stream of cheap labor than with the greed for cheap labor and the frantic development of industry which encouraged their migration. The responsibility is rooted in the sordid and senseless crowding of more and more industry into areas already overpopulated, where human dignity, decent morals, good diet and even a moderate education were and are virtually impossible.

We have been paying as a nation for the greed and planlessness of those urban monstrosities in terms of vice, of poverty, of malnutrition, of revolt and bitterness and foreign radical ideas and we have only begun to pay. The war has aggravated conditions already evil, and concentrated in already overcrowded areas millions more industrial workers, without economic security, who become grave social and economic problems to the community and the nation the moment factories begin to close and a depression threatens.

Today outside smaller and middle-sized industrial towns there are growing circles of small houses each with a small piece of land where industrial and white-collar workers have settled in order to find some degree of economic security and a better life for their families—better food, fresh air, better schools, better moral standards and lowered living costs. The dollar of the industrial worker in our great urban areas is worth only sixty to seventy cents as compared to the full value dollar of workers living in smaller communities. The worker in the great industrial centers spends a large portion of his dollar for higher real estate taxes, distribution costs and the higher overhead expenses which are a part of his economic surroundings. The consequent clamor is for higher industrial wages which in turn are passed on to the consumer who is the farmer, the serviceman, the middleman, and who in turn raises the prices of what they produce or the services they render, and within a period of months the industrial worker is back exactly where he started in so far as living costs are concerned. Then a new clamor arises for higher wages and the vicious circle begins all over again, a circle which in spiraling steadily upward keeps mounting until the disaster of unemployment causes the whole of our national economy to collapse. Certainly much of the violence of our economic depressions is created in our evil and overcrowded industrial cities.

The process of decentralization is one of the surest steps toward our economic security as a nation. The Tennessee Valley Authority has shown the way in industrial and power developments which permit workers to live sane and decent lives with "one foot on the soil." Industry itself is displaying a widespread interest in decentralization, in building new factories not in the old sordid, overcrowded cities, but in smaller towns and areas where the worker can earn a dollar worth one hundred cents and own a house and piece of land which gives him a stake in the nation and in democracy itself. In some small-town industrial areas countless workers today not only own and operate small holdings from one acre upward, but even farms of considerable size, an enterprise which the eight-hour day, the forty-hour week and modern compact and low-price farm machinery make possible.

One great Detroit industry, planning large expansion after the war, has already spread the manufacture of the implements it makes over more than one hundred small towns and communities rather than center all of it in a vast overcrowded industrial center like Detroit. Previous experiments have shown that smaller factories in smaller communities where workingmen can have a decent life and some degree of economic security are successful from the point of view of better working conditions, better health, less labor trouble, infinitely greater economic security.

The middle class is the backbone of democracy—in fact democracy cannot exist without a flourishing middle class. Perhaps the simplest definition of the middle class is that of a group of citizens who own something, who have some stake in individuality, in freedom, in good government, in the protection of civil rights and in the nation as a whole. Democracy is essentially a giant co-operative in which all the citizens have a stake. The middle class is the strongest bulwark against any totalitarian form of government. That is why it is always the first victim of totalitarian government, whether Fascist or Communist. A man with a stake in the nation is independent. He resists being pushed about and being regimented. A man without economic security, dependent upon the state to care for him whether it be to provide jobs or to pay him a dole when he is out of a job, is helpless. He can only continue to vote for the kind of government which provides him with a roof over his head, a miserable wage and food for the mouths of himself and his children. For him there is no security and no other way out.

I am aware of the virtues of unemployment insurance and of the social security measures established during the past few years but neither of these has been sufficiently extended to provide universal security and neither is, properly speaking, in the idiom of genuine democracy; nor does either provide that stake in government and the nation which makes for the democracy in which Thomas Jefferson believed so profoundly. Both are essentially makeshifts and both tend toward the increase of an expanding and costly bureaucracy.

There was a time when we had in this country no such thing as a proletariat. The very word is one that is alien to America, a word coined to fit the mass of dispossessed peasants and workingmen existing in backward European countries. Today we have a large and growing and very self-conscious proletariat, equally dispossessed and equally dependent upon government alone to guarantee them economic security. This proletariat, as the election returns of 1944 so clearly showed, centered in our great industrial cities and it is growing constantly in size.

Much of the bitterness and abuse aimed at the Political Action Committee of the C.I.O. during the campaign of 1944 should have been directed not at the committee but at the conditions which made the committee an inevitability. Much of the abuse was uttered by the very elements responsible for the conditions in our great industrial cities which have produced such a manifestation of the European proletariat—political action. The Political Action Committee is a symptom, a result, but not an activating cause. It is a manifestation of the unrest and insecurity of the rapidly growing class of economically insecure workers in our crowded industrial areas.

There has been much talk and many bitter words exchanged over "Socialistic" and "Communist" government measures established or proposed during the reign of the New Deal. Essentially all these measures have been taken or proposed to provide jobs, or relief, or subsidies for the growing number of dispossessed agricultural populations or industrial workers crowded into the cities. That population is growing and will be increased after the war by hundreds of thousands of rural workers who have left agricultural areas for the city. In time of unemployment or depression millions of this increasing group of citizens, devoid of economic security must be supported either by relief or make-work projects paid for out of higher and higher taxes from those elements of our society possessing economic security or the capital which provides projects which create employment. As the taxes increase, the economic sources which provide employment become more and more diminished and a vicious circle is created for which the only solution apparent under existing conditions and measures is greater and greater government control, government ownership, and totalitarianism. This progression may please some "Communists" and some "Socialists." They may even seek to promote it, but they are not responsible for it. The reasons are much more profound. They lie in growing lack of economic security for the individual, his growing lack of any stake in the nation other than a daily wage dependent upon prosperity, in increased migrations into industrial and urban centers, in the slow destruction of the solid and thrifty middle class without which a democracy cannot exist. The real problem, the real pressure arises from a growing proletariat, dispossessed of all property, and dependent upon the heavily taxed and waning economic strength of private ownership and initiative. That means simply a growing population *with votes* which expects to be cared for by the state. It can presently and quite simply vote democracy out of existence.

It is time, I think, to consider fundamental measures rather than to make laws and issue executive decrees, day by day, to meet emergency after emergency as it arises. Such a course is no more than drifting toward a state-owned and managed nation which few Americans, even those without any stake in the nation, welcome or will enjoy when it arrives.

These, I am aware, are tall thoughts. I think they are borne out by the evidence of history and by what I have seen in other nations.

"MY NINETY ACRES"

I HAD A FRIEND, A LITTLE OLD MAN, WHO LIVED OVER THE hill in Possum Run Valley in a small white house on a farm which is known as "My Ninety Acres." It has never been given that name as farms are named "Long View" or "Shady Grove." The name is not painted on the red barn nor on a fancy sign hanging at the end of the lane leading up to the house; nevertheless throughout the Valley everybody always refers to Walter Oakes's farm as "My Ninety Acres." At first, years ago when Walter was still a young and vigorous man, they used to speak of "My Ninety Acres" with a half mocking, half affectionate smile, especially the big farmers who owned a lot of land, because Walter always talked about that ninety acres as if it were a ranch of many thousand acres like the vast King Ranch in Texas, or a whole empire, as if he were Augustus Caesar or Napoleon referring to "My Empire." Some of the old farmers, I think, believed Walter a bumptious and pretentious young man.

But at last as time passed, and Walter turned into a solid middle-aged farmer and later into an old man, the smiles and mild sense of mockery went out and "My Ninety Acres" became simply the name of the place the way a farm was known as the Ferguson place or the Anson place. People said, "I'm going over to 'My Ninety Acres' " or "If you want to see a nice farm, go and have a look at 'My Ninety Acres.'" Nobody in the Valley any longer finds anything confusing or absurd about the name. I think this is so partly because in places like the Valley, people come to accept the name that is natural to a place and partly because as the years passed old Walter earned the right to say "My Ninety Acres" as Augustus Caesar might say "My Empire."

He had a right to speak of it with pride. It wasn't the conventional Currier and Ives farm one expects from the long tradition of American farming—a bright, new place, with new wire fences, and cattle standing like wooden animals in a pasture that was more like a lawn than a pasture. There was, indeed, a certain shagginess about it, a certain wild and beautiful look with that kind of ordered romantic beauty which was achieved by the landscape artists of the eighteenth century who fell under the influence of Jean Jacques Rousseau's romantic ideas regarding Nature. The white house was small but always well painted and prosperous in appearance, and there was no finer barn than Walter's with its fire-red paint, its big straw shed and its ornate shutters and cupolas painted white and there were no finer cattle in the whole county than those which stood behind the white-painted wooden fences of the barnyard staring at you, fat and sleek and contented, as you drove past "My Ninety Acres."

The romantic shagginess appeared too in the garden around the small white house with its green shutters that stood beneath two ancient Norway spruces. The patches of lawn were kept neatly mowed but surrounding them grew a jungle of old-fashioned flowers and shrubs—lilacs, standing honeysuckle, syringa, bleeding heart, iris, peonies, tiger lilies, day lilies, old-fashioned roses like the Seven Sisters and the piebald and the Baltimore Belle. At the back the little vegetable garden was neat enough with its rows of vegetables and its peach and pear and quince trees in a row inside the white picket fence. But beyond the borders of the garden, the shagginess continued. There weren't any bright, new, clean wire fences. The wire along the fence rows was hidden beneath sassafras and elderberry and wild black raspberry and the wood lot on the hill above the creek was not a clean place with the grass eaten short by cattle. The cattle had been fenced out and the trees, from seedlings to great oaks grew rankly with a tropical luxuriance.

But despite the shagginess of the farm's appearance no fields in the Valley produced such big crops or pastured such fine cattle and hogs. At "My Ninety Acres" the shagginess didn't exist, the neighbors came to understand, because Walter was lazy or a bad farmer—there was no more hard-working man in the whole Valley. They were that way because Walter wanted them like that—Walter and Nellie.

I never saw Nellie Oakes. She died before I was born, but my father told me about her. In his time she had been the prettiest girl in the Valley and she taught school at the Zion School house until, when she was twenty-two, she married Walter Oakes. People wondered why she chose him when she might have married Homer Drake whose father owned four hundred and fifty acres of the best land in the county or Jim Neilson whose family owned the bank and the feed mill in Darlingtown. She could have had her choice of any of the catches of the Valley and she chose Walter Oakes, who had no more than ninety acres of poor hill land he had just bought because he didn't have money enough for anything better.

In the parlor of the little white house on "My Ninety Acres" there hangs an old enlarged photograph of Walter and Nellie taken at the time of their marriage. It is hand-colored and the bride and bridegroom are standing like statues, each with a clamp obviously fastened at the back of the heads in order to "hold the pose," but even the stiffness and artificial coloring cannot alter or subdue the look of youth and health and courage that is in both of them. Walter, the thin, tough old man who was my neighbor and friend, stands there in the photograph, stalwart and handsome and full of courage, one big muscular hand on Nellie's shoulder. He was blond with blue eyes and the gentle look which big, strong men often have because there is no need for them to be pugnacious or aggressive.

On a chair, beside and a little in front of him, sits Nellie in a white dress with leg-o'-mutton sleeves and a full flounced skirt—dark, more beautiful than pretty—with big dark eyes, holding in her small hands a lace handkerchief and a bunch of lilacs. I think Nellie was beautiful rather than pretty because of the look of intelligence. Even today, you sometimes hear old people say, in the Valley, "Nellie Oakes was a mighty smart girl—the only woman I ever knew who was as smart as she was pretty."

Nellie, so far as I can discover, never told anybody why she chose to marry Walter instead of one of the catches of the Valley, but I know from all the long story that it was because she was in love with him. As it has turned out, she was right because the big four hundred and fifty acre Drake place which Homer inherited has gone downhill ever since Homer took possession of it and today, with its worn-out fields and decaying buildings, it wouldn't bring as much as "My Ninety Acres" and Jim Neilson died long ago as a drunkard, having lost both the bank and the feed mill. But "My Ninety Acres" is the richest, prettiest farm in all the county, although Nellie isn't there to enjoy its beauty and prosperity. I say she isn't there because she died a very long time ago. But sometimes when I walked about the fields of "My Ninety Acres" with old Walter, I wasn't at all sure she wasn't there, enjoying its beauty and richness as much as old Walter himself.

I am forty-eight years old and Nellie died before I was born when she gave birth to her second son, Robert.

My father was a gentle man. He never went through the Valley without stopping at "My Ninety Acres" and usually I was with him. Sometimes when we stopped at "My Ninety Acres" for a meal or for the night, I stayed and played about the barn with Robert Oakes who was two years older than I and his brother John who was two years older than Robert. Sometimes if it was a Sunday we went fishing or swimming. Sometimes I simply trudged behind my father and Walter Oakes and his two sheep dogs as they walked about "My Ninety Acres," and as I grew a little older I sometimes wondered that the two men could be together, walking side by side, perfectly happy, without talking at all. I did not know then what I came to know later that among men who were as close to each other as my father and Walter Oakes, conversation wasn't necessary. They knew without speaking what the other felt when a lazy possum, out in the middle of the day when he shouldn't have been, lumbered across the pasture and out of sight and scent of the dogs (I've seen Walter call the dogs and keep them by his side till the possum had disappeared, safe in some deep hole or hollow log).

And I was always a little surprised at how often Walter would say, "Nellie wanted me to put this field into pasture but we couldn't afford not to use it for row crops," or "'Nellie was smart about such things," or "It's funny how many good ideas a woman can have about farming. Now, Nellie always said. . . ." Sometimes in the warm summer heat, I'd return to the house, still trudging along behind the two grown men and the dogs, believing that I would find there the Nellie whom I had never seen, who was dead before I was born, waiting for us with a good supper on the table.

But Nellie was never there. There was only an elderly widow woman called Mrs. Ince, a distant cousin of Walter's who came to keep house and look after him and the boys after Nellie Oakes died. She was a queer old woman, very thin and very active, who was always asking Walter how Nellie had molded the butter or pickled the beets or kept a broody hen on the nest because she wanted everything to be the way Walter liked it. She could not have been more than fifty for she was still young enough to create talk in the Valley about her living there alone on "My Ninety Acres" with Walter and the boys, but to a small boy like myself she seemed immensely old. She was, as I remember her, very plain and kind and dull with the meekness which often characterized indigent widows of her generation who were grateful for a roof over their heads, something to eat and a little spending money. When she came to "My Ninety Acres" some of the old women in the Valley talked of the impropriety of her living there in the same house with Walter. I know now that anyone who had ever known Nellie must have been mad to think that Walter Oakes ever had any thoughts about poor, drab Mrs. Ince. She was at most, a convenience, someone to do the cooking and baking and housekeeping for a vigorous man and two wild, vigorous boys.

People in the Valley couldn't see why Walter Oakes didn't get married again. They said, "He's still a young man and he's done a wonderful job with 'My Ninety Acres,'" or "I don't see how a man like that can get on without a woman at his age. It ain't natural." And a good many widows and spinsters past their first youth certainly set their caps for him. It wasn't only that he was doing well with "My Ninety Acres," he was, as I remember him then, a big, straight, clean good-looking fellow with his sun-tanned face, blue eyes and blond hair bleached by the sun. He would, I think, have pleased even a young girl.

But Walter never showed any signs of marrying again. He was always polite and his eyes sometimes twinkled with humor when he saw what some of the good ladies were up to. He didn't leave "My Ninety Acres" save to go into town to buy or sell something or to go to the Valley church on Sunday with Mrs. Ince and the boys. He'd come home from church and change his clothes and spend the rest of the day walking round the place. Sometimes, to the scandal of the old ladies of the Valley, he'd plow or make hay with the boys on a Sunday afternoon. I remember him saying to my father, "They talk about my working on Sunday or plowing, but when the ground is ready or hay has to be taken in, it has to be taken care of. The good Lord wouldn't like to see his beasts eating poor hay all winter because some old woman said it was wrong to work on Sunday. Nellie always said 'The better the day, the better the deed' and quoted that bit of the Bible about the ox falling into the ditch."

The two boys were nice kids and smart like Nellie. John, the older one, looked like her, with dark eyes and dark hair. Robert, the younger one, who had never seen his mother, looked like Walter. The father wanted both of them to go to college and get a good education. I think Walter always loved John, the older one, best—not because of any resentment of Robert because he had caused his mother's death but because John looked so much like Nellie.

With all my family, I went away from the county when I was seventeen and I was gone for twenty-five years. Sometimes at first my father heard from Walter, rather brief, unsatisfactory and inarticulate letters, written on lined paper torn out of a copybook, but neither Walter nor my father were very good letter writers. They were both the kind of men who could not communicate without the warmth that came of physical presence. Writing letters didn't mean much. When they met again, even after years, the relationship would be exactly the same. They were that kind of men, and that kind of friends.

I know very little of the details of what happened during those years, only a fact or two and what little I have picked up from Walter as an old man in his implications regarding the past. The war came and in it John, the older son, whom Walter secretly loved best, was killed at St. Mihiel. He was twenty-one and just finished with agricultural college. Walter had counted on his returning to the farm, marrying and producing grandchildren to carry it on. Robert when he returned from the war, did not stay on the farm. He was very smart, like Nellie, but he didn't want to be a farmer.

Robert had ambitions. He had had them even as a small boy. Sometimes when the three of us, as kids, sat naked among the wild mint by the swimming hole, we talked about what we were going to do in life and Robert always said, "I'm going to be a great man and get rich and have an automobile with a man to drive it."

In the twenty-five years I was away from the Valley, Robert had achieved exactly what he had planned. By the time I returned to the Valley, Robert was president of the Consolidated Metals Corporation and he had made many millions of dollars. I think he must have had both Nellie's "smartness" and Walter's steadfastness.

In the first weeks after I came home, I never thought about my father's friend, old Walter Oakes. Indeed, I had very nearly forgotten his existence. And then one day I heard Wayne, one of the boys on the farm, say something about "My Ninety Acres" and I remembered it all and asked, "Is Walter Oakes still alive?"

"Alive!" said Wayne, "I'll say he's alive. The livest old man in the county. You ought to see that place. Brother, that's the kind of farm I'd like to own. He raises as much on it as most fellows raise on five times that much land."

Wayne, of course, was only twenty. He couldn't know how once people had laughed when Walter Oakes spoke proudly of "My Ninety Acres." Clearly, they didn't laugh any more. Clearly, Walter Oakes was the best farmer in all the county, very likely the best farmer in all the rich Ohio country.

The next Sunday I walked over the hills to "My Ninety Acres." As I came down the long hill above the farm I saw that it hadn't changed much. The house still looked well-painted and neat with its white walls and green shutters and the barn was a bright new prosperous red. But the shrubs and flowers had grown so high that they almost hid the house. It was a day in June and as I walked down the long hill the herd of fat, white-faced cattle stood knee-deep in alfalfa watching me. I hadn't taken the dogs because I knew Walter always kept a couple of sheep dogs and I didn't want a fight.

As I walked down the hill I thought, "This is the most beautiful farm in America—the most beautiful, rich farm in the world—'My Ninety Acres.'"

The corn stood waist-high and vigorous and green, the oats thick and strong, the wheat already turning a golden yellow. In the meadow the bumblebees were working on clover that rose almost as high as a man's thighs. In all that plenty there was something almost extravagant and voluptuous. The rich fields were like one of the opulent women painted by Rubens, like a woman well loved whose beauty thrives and increases by love-making.

I pushed open the little gate and walked into the dooryard with the neatly mown grass bordered by lilacs and peonies and day lilies. The door stood open but no one answered my knock and thinking the old man might be having a Sunday nap, I stepped into the house and called out, "Walter! Walter Oakes!" But no one answered me.

I hadn't been in the house for twenty-five years and I didn't remember very well my way about it so when I opened the door which I thought led into the long room that had once been used both for eating and living, I found that I was mistaken. I had stepped into the parlor instead.

It had that musty smell of country parlors and the shutters were closed but there was enough light for me to see the enlarged hand-colored portrait of Walter Oakes and his bride Nellie hanging on the wall above the fireplace. Out of the stiff old picture they looked at me young, vigorous, filled with courage and hope and love. It struck me again how pretty Nellie was.

I stood for a little time looking at it and then turned and closed the door behind me. I went out through the sitting room and the kitchen where everything looked clean and neat as in the dooryard, and I thought, "He must have a woman to look after him."

By now, of course, I remembered enough to know that I should find old Walter somewhere in the fields. Sunday afternoon he always spent walking over the place. As a small boy I had followed him and my father many times.

So I went down toward the creek and as I turned the corner by the barnyard I saw him down below moving along a fence row. Two sheep dogs were with him, the great-great-great grandchildren of the pair I had known as a boy. They were running in and out of the hedgerow yapping joyously. I stood for a moment, watching the scene. The fence row bordered a meadow of deep thick hay and below among feathery willows wound the clear spring stream where I had often gone swimming with Walter's boys—John who had been everything Walter had hoped for in a son, the best loved, who was buried somewhere in the Argonne, and Robert who had gone away to become rich and powerful. There was something lonely about the figure of the old man wandering along the fence row filled with sassafras and elderberry. For no reason I could understand I felt a lump come into my throat.

Then I noticed that there was something erratic in the progress of the old man. He would walk a little way and then stop and, parting the bushes, peer into the tangled fence row. Once he got down on his knees and for a long time disappeared completely in the thick clover.

Finally, as he started back along the far side of the field, I set off down the slope toward him. It was the barking of the dogs as they came toward me that attracted his attention. He stopped and peered in my direction shading his eyes with his big hands. He was still tall and strong, although he must have been well over seventy, and only a little stooped. He stood thus until I was quite near him and then I saw a twinkle come into the bright blue eyes.

"I know," he said, holding out his hand. "You're Charley Bromfield's boy. I heard you'd come back."

I said I'd been trying to get over to see him and then he asked, "And your father? How's he?"

I told him my father was dead. "I'm sorry," he said, very casually as if the fact of death were nothing. "I hadn't heard. I don't get around much." I explained that my father had been ill for a long time and that death had come as a release.

"He was a good man," he said. "A fine man. We sort of dropped out of writing each other a good many years ago." He sighed, "But after all writing don't mean much." The implication of the speech was clearly enough that friends communicated without writing, no matter how great the distance that separated them.

Then suddenly he seemed to realize that I must have seen him for a long time, ducking and dodging in and out of the fence row. A faint tinge of color came into his face and he said shyly, "I was just snoopin' around my ninety acres. I like to see what goes on here and I don't get time during the week."

He looked down at his big hands and noticed, as I did, that some of the black damp loam of the fence row still clung to them. He brushed them awkwardly together. "I was just digging into the fence row to see what was going on there underground. A fellow can learn a lot by watching his own land and what goes on in it and on it. My son John—you remember the one that was killed in the war—he went to agricultural school but I don't think he learned more there than I've learned just out of studying my own ninety acres. Nellie always said a farm could teach you more than you could teach it, if you just kept your eyes open . . . Nellie . . . that was my wife."

"Of course," I said, "I remember."

Then he said, "Come with me and I'll show you something."

I followed him along the fence row and presently he knelt and parted the bushes and beckoned to me. I knelt beside him and he pointed, "Look!" he said, and his voice grew suddenly warm, "Look at the little devils."

I looked and could see nothing at all but dried brown leaves with a few delicate fern fronds thrusting through them. Old Walter chuckled and said, "Can't see 'em, can you? Look, over there just by that hole in the stump." I looked and then slowly I saw what he was pointing at. They sat in a little circle in a tiny nest, none of them much bigger than the end of one of old Walter's big thumbs—seven tiny quail. They sat very still not moving a feather, lost among the dry, brown leaves. I might not have seen them at all but for the brightness of their little eyes.

"Smart!" he said, with the same note of tenderness in his voice. "They know! They don't move!"

Then a cry of "Bob White!" came from the thick, fragrant clover behind us and Walter said, "The old man's somewhere around." The whistle was repeated, again and then again.

Old Walter stood up and said, "They used to laugh at me for letting the bushes grow up in my fence rows, but they don't any more. When the chinch bugs come along all ready to eat up my corn, these little fellows will take care of 'em." He chuckled, "There's nothing a quail likes as much as a chinch bug. Last year Henry Talbot, down the road, lost ten acres of corn all taken by the bugs. Henry's a nut for clear fence rows. He doesn't leave enough cover along 'em for a grasshopper. He thinks that's good farming, the old fool!" and the old man chuckled again.

We were walking now up the slope from the creek toward the house, and he went on talking, "That fence row beside you," he said, "is just full of birds—quail and song sparrows and thrushes—the farmers' best protection. It was Nellie that had that idea about lettin' fence rows grow up. I didn't believe her at first. I was just as dumb as most other farmers.

But I always found out that Nellie was pretty right about farmin'. She was hardly ever wrong . . . I guess never."

As we reached the house, old Walter said, "Funny how I knew you. I'd have known you anywhere. You're so like your father. I've missed him all these years, especially when anything happened he would have liked . . ." he chuckled, "like these baby quail today. Come in and we'll have a glass of buttermilk. It's cooler in the sittin' room."

I went with him into the springhouse. It was built of stone with great troughs inside cut out of big blocks of sandstone and the water ran icy cold out of a tile that came through the wall. Cream, milk and buttermilk, stood in crocks in the icy water, each covered by a lid held in place by an ancient brick with velvety green moss growing on its surface. Coming out of the heat into that damp cool spot was like coming into another world.

He picked up a pitcher with buttermilk in it and I asked, "Who does your churning for you?"

He grinned, "I do it myself," he said. "Of an evening. I kinda like it."

We went and sat in the living room and he brought glasses and two white napkins. It was buttermilk such as I had not tasted in thirty years—creamy, icy cold with little flakes of butter in it.

I said, "What became of Mrs. Ince?"

He said, "Oh, she got old and sick and went back to live with her sister. I just didn't get anybody to take her place."

"You mean you're living here all alone?" I asked.

"Yes."

I started to say something and then held my tongue, but old Walter divined what it was I meant to ask and said, "No. It ain't lonely. I've always got the dogs. Jed Hulbert comes down and helps me with jobs I can't do alone and his wife takes care of my laundry and cleans up once a week. Jed and his wife like the money and they're nice people." He smiled, "It doesn't seem to me like a farm is a lonely place. There's too much goin' on. Nellie used to say she didn't understand the talk of these women who said they got lonely. Nellie said there was always calves and horses and dogs and lambs and pigs and that their company was about as good as most of them women who talked that way. And she always had her posy garden. Did you notice it coming in? It's mighty pretty right now. Nellie planted everything in it . . . just the way they are today." He was about to say something else but checked himself and looked at me strangely. A secretive, almost sly look came into his eyes and he turned away to stare at the glass he held in his hand.

After an awkward pause I said, "Well, Robert did all right by himself. He always said he wanted a big automobile and a driver and a lot of money and he got it all right."

Then old Walter looked up at me and grinned, "Yes, I guess he got just about what he wanted. He's a good boy, but he's got some funny ideas." The old man chuckled. "He's been trying for years to get me to retire and live in the city where I could take it easy or go down and live in Florida. What'd I do with these big ugly hands in a place like that? I wouldn't know what to do with myself. And what would become of 'My Ninety Acres'?" Or he's always wantin' to buy me a bigger place with a house full of gadgets or to buy me a lot of machinery. What would I want with a bigger place? Ninety acres is enough for any man if he takes care of it right, like he should. And anyway it wouldn't be the same as "My Ninety Acres." And I don't want machinery bought with his money. 'My Ninety Acres' ought to buy its own machinery and it does." A fierce note of pride came into his voice. "All the machinery it needs. Robert wants me to hire a couple to live here and do the work for me, but I wouldn't like that. Yes, Robert's got some crazy ideas and he doesn't understand how I feel. I guess he thinks I'm a little crazy."

It was getting late and I rose, but the old man went on talking. "It's a pity about Robert not having any children. I guess his wife is all right. I don't see much of her. We don't have much in common. But it's a pity Robert couldn't have found a woman he could have loved."

That was the first and last time I ever heard him speak of his daughter-in-law, but out of the meager speech and the look in his eyes and the sound of his voice I divined what she must be like. Indeed, I gained a very clear picture of her.

"Robert comes to see me about once a year and stays for a day or two, but he's a pretty busy man with all the big affairs he has to manage."

"Tell him to drive over and see me the next time he comes," I said, "And you come over too."

He opened the screen door for me. "I'm afraid I don't get off 'My Ninety Acres' very often any more. You'll understand if I don't get over soon. The place takes a lot of time when you're working it alone."

I left him and the dogs at the gate and set out over the hill across the pasture with the fat, white-faced cattle, for home.

It wasn't the last time I saw old Walter. There was enough of my father in me to make the friendship between myself and the old man before long very nearly as warm as their friendship had been. And after all, between them, they had taught me many of the things I had come with experience to value most in life. The Sunday afternoon visits to "My Ninety Acres" became very nearly a habit, for I found gradually that old Walter was in himself an education. He knew more of the fundamentals of soil, of crops, of livestock than any man I have ever known. Some of them he had read in books and in farm papers but he didn't trust the things he read until he tried them out and many of them he didn't even attempt to try out since out of his own wisdom he understood at once that they were rubbish. Instinctively and out of experience he rejected things which ran counter to the laws of Nature.

"Nellie," he would say, "always said that Nature and the land itself was the best answer to all these questions. If it wasn't natural it wasn't right, Nellie would say, and I've never found that she was wrong. She used to say that there were two kinds of farms—the 'live' farms and the 'dead' ones and you could tell the difference by looking at them. A 'live' farm was the most beautiful place in the world and a 'dead' farm was the saddest. It depended on the man who worked them—whether he loved the place and saw what was going on or whether he just went on pushing implements through the ground to make money. Nellie was awful smart about a lot of things."

Sunday after Sunday we would make a round of the small empire while old Walter told me the history of each field and what had happened to it, what he had learned from this field or that one, and why his alfalfa and clover were thicker than those of his neighbors, his corn higher and sturdier, his Herefords bigger and fatter. And after a time I began to understand how old Walter and my father could walk side by side half the afternoon without speaking to each other, communicating by a smile or a nod or without any visible or audible sign. There are times when speech is a poor, inadequate business.

One afternoon I arrived to find old Walter in the garden, standing quite still, staring at something. He did not speak when I came near him but only raised his hand in a gesture which clearly prohibited any speech or violent movement. Then he pointed at a male cardinal, very handsome in his red coat, moving restlessly about the lower branches of a magnolia and chirping anxiously. In a low voice he said, "The poor fellow is looking for his mate. I found her dead yesterday on the ground under that pine over there. He was staying around, trying to bring her back to life and make her fly away with him. I took her and buried her. I hoped he'd forget and fly away and find another mate. But he didn't. He keeps hanging around, trying to find her. It's funny about birds and animals that way."

Then a farmer and his wife came in the gate and interrupted our quiet. We were not always alone on those Sunday walks because neighbors and even farmers from a great distance came sometimes on Sundays to see his farm and hear him talk about "My Ninety Acres." I knew he took pride in his prestige but he never showed it. He kept his simple, modest manner when he talked of this field or that one, a kind of fire would come into the blue eyes, like the fire in the eyes of a man talking of a woman he loves passionately. He never came to see me but he always welcomed me warmly on "My Ninety Acres" and when I missed a Sunday, he was disappointed.

And then one brilliant day in October I saw a big, shiny black car coming up the long lane to our house. I knew at once who was in it. I knew by the size and importance of the car, and as it drew nearer, by the cut of the driver's uniform. It was Robert. He had come on his annual visit and had driven over to see me.

I went down the path to meet him and as he stepped out of the shiny car, it was hard for me to remember him as the boy I had seen the last time when he was sixteen, slim, muscular, towheaded and athletic. He still looked a little like old Walter, yet in a strange way he appeared older than the old man. He was plump and rather flabby with pouches beneath the eyes which looked through the shining lenses of steel-rimmed spectacles. He stooped a little and there was a certain softness about his chin and throat.

He said, "I'm Robert Oakes. My father told me you had come back to live in the Valley."

"Yes, I know. I'm delighted to see you. Come in."

I found him rather as I had expected him to be, an intelligent fellow, with a good deal of dignity and authority. He was, after all, the child of old Walter and Nellie and their qualities could not be altogether lost in him. After thirty years the going was a little stiff at first but after a drink we got together again, mostly by talking about "My Ninety Acres" and the old swimming hole in the creek and maple sugar making time and the other boyhood experiences we had shared.

He laughed once and said, "The old gentleman has certainly made good on his ninety acres."

I asked him to stay for lunch and he accepted the invitation so readily that I suspected he had counted on it from the beginning. I said, "I know it's no good sending for your father. He won't leave the place."

"No, he and Jed were in the field by the creek husking corn when I left." Robert laughed, "He told me if I sat around long enough over here I'd get a drink and be asked to lunch. He said it was worth it to see the house and the place. Privately, I think he wanted to get rid of me for most of the day so he could get on with his work. He doesn't know what to do with me. I get in his way and take up his time."

We had lunch at a table crowded with noisy children with four dogs on the floor beside us. I think, at first, that Robert didn't know quite how to take it, but he warmed up presently and said to me, "You have a mighty good life here. I envy you."

After lunch we sat for a time on the porch overlooking the Valley. The sky was the brilliant blue of an Ohio sky in October and the trees were red and gold and purple with the green winter wheat springing into life in the fields beyond the bottom pasture where the Guernseys moved slowly across the bluegrass. He kept watching the Valley, so intently at times that he did not seem to hear what I was saying.

And presently he came round to what was clearly the object of his visit. "I really wanted to talk about my father," he said. "He's quite a problem and stubborn as a mule. I know your father was a great friend of his and that he accepts you nowadays exactly as if you were your father. And I thought you might have some influence with him. You see, I offered him almost everything—I've offered him a fruit ranch in Florida or Southern California, or a bigger farm, or a flat in New York. I've tried everything and he doesn't want any of it. He won't even let me hire him a couple or buy him an automobile or any machinery that might make life easier for him. This morning he was up at daylight and down husking corn in the bottom field with Jed by seven o'clock."

I grinned for I could see the whole picture and could understand how the old man's rich, famous, successful son, got in his way.

"When I got up," said Robert, "I found some eggs and pancake batter laid out for me and coffee on the stove, with a note to my driver about how to get breakfast for me. In the note he said to come down to the bottom when I'd finished breakfast. What can you do with a fellow like that?"

"What do you want me to do?"

"I want you to persuade him to let me do something for him. He's seventy-five years old and I'm afraid something will happen to him alone there in the house or barn."

"I'm afraid it's no good," I said. "I couldn't persuade him any more than you."

"I've tried everything even to saying 'What would it look like if it came out in the papers that my father had died suddenly alone on his farm in Ohio?' That's pretty cheap, but even that didn't move him. All he said was, 'You're rich enough to keep it out of the papers and anyway the dogs would let people know if I was sick.'"

We were both silent for a time and then I said, "Honestly, Bob, I don't think there's anything to be done and to tell the truth I don't see why we should do anything. He's as happy as it's possible for a man to be. He's tough as nails and he loves that place like a woman." Then hesitantly, I said, "Besides, Nellie is always there looking after him."

A startled look came into the son's blue eyes and after a moment he asked, "Do you feel that way, too?" Nellie, who died when Robert was born, must have been as unknown and strange to Robert as she was to me.

I said, "I think Nellie is everywhere in that ninety acres. He's never lonely. She's in the garden and the fields and his famous fence rows. She's out there husking corn with him now in the bottom forty."

Robert lighted another cigar. "It's the damndest thing," he said. "Sometimes I've felt that he had some resentment because I killed my mother when I was born or that he liked John better because he looked like her, but I know that isn't true. That's not in the old gentleman's character. I think it's more because Nellie is always there and I just get in his way. It's funny," he added, "I always think of her as Nellie—somebody I would have liked knowing because she was so pretty and kind and gay and 'smart' as they say here in the Valley. Sometimes I think the old gentleman gets Nellie and the ninety acres a little mixed up."

We talked some more and then Robert called his driver, got in the shiny car and drove off. We had agreed that there wasn't anything to be done about old Walter and Nellie. I said I'd keep my eye on him and go over myself or send somebody once every day to see that he was all right. Of course on Thursdays it wasn't necessary because that was the day that Jed's wife came to do the washing and clean up. And so every day for two years I, or somebody from the place, went over. Sometimes we'd have an excuse but more often we didn't even let him know that he was being watched. One of us would drive past at chore time, or I'd walk over the hills and watch until he appeared in the barnyard or the garden. I knew how much he'd resent it if he suspected that anyone was spying on him, and I didn't want to risk breaking our friendship.

I continued to go over every Sunday and each time I went over I learned something about soil, or crops or animals, for the knowledge and experience of the old man seemed inexhaustible. And then one Sunday afternoon in early September when we were walking alone through one of old Walter's cornfields, I made a discovery. It was fine corn, the whole field, the best in the whole county, and as we came near the end of a long row, he stopped before a mighty single stalk of corn which was beautiful in the special way that only corn can be beautiful. It was dark green and vigorous and from it hung two huge nearly ripened ears and a third smaller one. Old Walter stopped and regarded it with a glowing look in his blue eyes.

"Look at that," he said. "Ain't it beautiful? That's your hybrid stuff." His hands ran over the stalk, the leaves and the ears. "I wish Nellie could have seen this hybrid corn. She wouldn't have believed it."

As I watched the big work-worn hand caressing that stalk of corn, I understood suddenly the whole story of Walter and Nellie and the ninety acres. Walter was old now, but he was vigorous and the rough hand that caressed that corn was the hand of a passionate lover. It was a hand that had caressed the body of a woman who had been loved as few women have ever been loved, so passionately and deeply and tenderly that there would never be another woman who could take her place. I felt again a sudden lump in my throat, for I knew that I had understood suddenly, forty years after the woman was dead, one of the most tragic but beautiful of all love stories. I knew now what Robert's strange remark about Nellie and the ninety acres getting mixed up had meant. Robert himself must once have seen something very like what I had just seen.

It happened at last. I went over one Sunday afternoon a few weeks later and when I could not find old Walter or the dogs anywhere I returned to the house and went inside. I called his name but no one answered and in a little while I heard scratching and whining in the ground floor bedroom and then a short, sharp bark and when I opened the door the sheep dog bitch came toward me. The other dog lay on the hooked rug beside the bed his head between his paws, looking at me mournfully as if he knew that I understood. On the bed lay old Walter. He had died quietly while he was asleep.

I telegraphed to Robert and he came with his wife for the funeral. The wife was exactly as I expected her to be and I understood what old Walter had meant when he said it was a pity Robert had never found a woman he could love. As I listened to the service, I knew how much feeling lay behind old Walter's simple observation.

He was buried beside Nellie in the Valley churchyard. The dogs came over to join my dogs and after awhile they got on together. Robert wouldn't sell "My Ninety Acres" but I undertook to farm it for him and one of our men went there to live. But it will never be farmed as old Walter farmed it. There isn't anybody who will ever farm that earth again as if it were the only woman he ever loved.

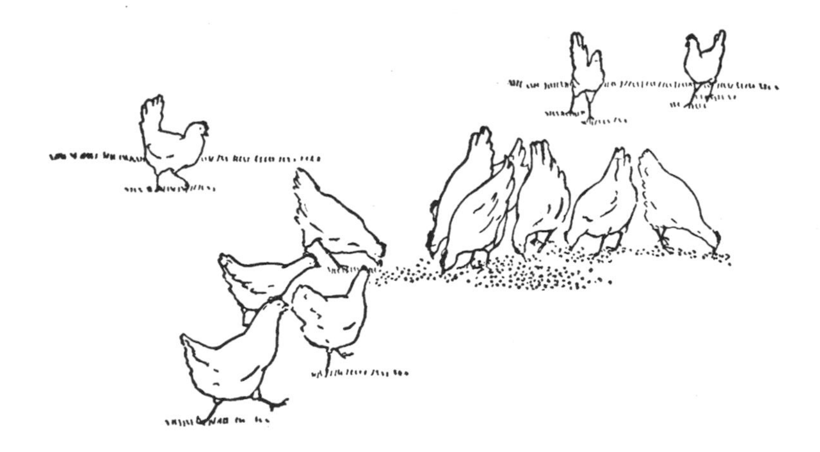

THE WORLD WITHIN THE EARTH

"The best fertilizer on any farm is the footsteps of the owner."
—Confucius

Like Nellie Oakes, I have a strong belief that there are both "living" farms and "dead" farms and that the difference lies not in the soil but in the men who own them and farm them. A drive along any country road will demonstrate this difference.

There is no more good virgin soil in this country and most of the rich virgin soil that existed in the beginning has been farmed long enough for it to take on the stamp and character of the men who have owned and farmed it. What I am trying to say is simply this—that if a good man or good men have farmed soil for a hundred years or more, it will still be rich and productive land. On the other hand, if shiftless greedy men have exploited it, there will be little left that is any longer worth anything either to the owner or to the nation.

The world of Pleasant Valley, like the world of most of the United States, has been farmed long enough for the reckoning to be made and the accounts partly settled. When I am asked what is the value of land in my part of the world, I can give no definite answer because the imprint of man, written upon the land, has altered its value. Once it was all good land, varying only a few dollars an acre according to whether it was bottom land or hill land, but today there is no rule. One farm, according to its buildings and productivity, may be worth two hundred dollars an acre and the farm next to it be worth only twenty or thirty. All the balance has been upset by the records of the men who owned or farmed the land. The imprint of what they have done is written there on the earth itself—their thrift, their intelligence, their energy, their "feel" for the earth, or their greed, their stupidity, their shiftlessness.

At Malabar we had alluvial bottom fields which are exhausted and steep hillside fields, like those on the Anson place that were highly productive. Clem Anson was a live farmer, who loved his hillside land; the men who farmed the lower fields were stupid men and blindly greedy, destroying their own capital day by day.

A good farmer must be many things—a horticulturist, a mechanic, a botanist, an ecologist, a veterinary, a biologist and any number of other things—but knowledge alone is not enough. There must be too that *feel* of all with which nature concerns herself. That is what Confucius meant in the proverb which heads this chapter. A good farmer, knows from day to day, even from hour to hour the state of the weather, of his crops, and his animals. A good farmer, a "live" farmer is not one who goes into the field simply to get the job of plowing completed because he must first turn over the soil in order to plant the crops that will bring him in a little money. The good, the "live" farmer is the man who turns now and then to look back at the furrows his plow turns over, to see that the soil crumbles behind him as rich, good soil should crumble. He is the man who sees the humus in his good earth and counts the earthworms, and watches his crops as they grow to see whether they are strong or sickly and what it is that is needed to make his pastures dark and green, his corn tall and strong. He is the man who learns *by farming*, to whom the very blades of grass and stalks of corn tell stories. He is the man to

whom good crops sing a song and poor ones convey a painful reproach. He is the man who knows that out of the soil comes everything, that out of the soil come the answers to the questions which torment him. He is, I think, the happiest of men for he inhabits a world that is filled with wonder and excitement over which he rules as a small god; it lies with him whether his world shall be rich and prosperous or decaying and poor and wretched. He is the man who knows how deeply Nature can reward the conscientious, intelligent worker and how bitterly she can punish the stupid shiftless ones to the very marrow of their bones.

Agriculture is in many ways the most satisfactory and inexhaustible of sciences. It touches or includes countless other fields of knowledge and research and with each discovery whole new fields to be explored are opened up. It is like the Pleasant Valley country where beyond each wooded hill there is a new small world, filled with adventure, mysterious and complete within itself. Now and then a professor, a bureau man or a scientist leads the way into the wrong country as did the man who believed that chemical fertilizer is the panacea for the ills of the soil. It took many years and millions of acres of soil made arid, bereft of humus and bacteria and earthworms for the farmer to discover that chemical fertilizer was not the *whole* answer but only a small, though vital, part. In the meanwhile many farmers and a good many scientists neglected the actual process of what went on in the soil, forgetting that the natural process of health and growth and reproduction is an immensely complicated affair, whether in man or in beast or in plant and that there are no shortcuts which are not in the end illusory and costly.

With all the research we have made there still remain many mysteries, not beyond explanation but which have not yet been explained or understood. In this borderland the "live" farmer finds his place—the man who sees and feels what is going on in the soil beneath his feet and on the earth around him, the man whose footsteps are the best fertilizer of his farm.

The best book I know on soil and the processes which take part in it is *An Agricultural Testament* by Sir Albert Howard (Oxford University Press). It is the record of a lifetime spent in working with soil, its preservation and restoration, and the author approaches the whole subject with a kind of reverence and mysticism without which I believe the efforts of any agricultural scientist may, in the end, be of little value. Possibly no one knows more of soil and the principles that effect growth and fertility and abundance than Sir Albert, but his attitude is one of humility toward the still greater mysteries which he has not yet explored.

A disciple of his and a friend of mine is the manager of one of the great stud farms of racing stock in this country and he was called to the job when the owner found that the quality of his colts was going rapidly downhill and that the colts broke down in the first few months of training. Everything known to race horse breeders was tried to build bone and stamina in the colts, for the stud farm and the racing stables were both acquiring a bad name, but the condition grew steadily worse, and as a last resort the owner summoned from England my friend, the disciple of Sir Albert Howard, who believes that the history of animals and plants and even of mankind is largely written in soil.

The soil of parts of Kentucky and of Ireland are famous for the quality of the horses bred and raced there and the reason has long been apparent. It is that the soil of both regions contains or did contain the proper amounts of calcium and phosphorus and trace minerals to produce strong bone and a maximum of stamina.

The stud farm which my friend took over lay in the very heart of the Kentucky horse-breeding area. Nevertheless when he went to the soil to find the secret of the stud's decline, he discovered that through farming over a long period of years in the violence of a climate in which there was freezing and thawing and heavy and violent rains, the amount of calcium and phosphorus in the land had become largely exhausted or had leached away. The soil held the clue. No amount of breeding could have changed the quality of the horses feeding upon that depleted soil. Once calcium and phosphorus were returned to the soil of that farm, the quality of the colts produced there rose again to the old standards of a racing stable which once had been famous.

The same friend is a "live" farmer who sees what is going on under his feet. He made the discovery that the grass in one paddock reserved for brood mares over a period of years, was finer and lusher than in the good pastures adjoining, and the discovery opened up a whole world of speculation regarding the effect of hormones and the glandular secretions of animals upon soil. Pushing the discovery a little further and making a careful check in a second paddock, he discovered that the urine of mares in season had an extraordinarily vitalizing effect upon the bluegrass of that paddock.

A heap of well-cured barnyard manure is not merely a composite of certain elements held together in certain chemical formulas. It is not merely the residue of calcium, phosphorus, potassium, nitrogen and trace minerals which the animals have eaten, rejected and passed out of the intestines. There are other mysterious elements—many kinds of bacteria, hormones, vitamins, glandular secretions—of which we know all too little. You could make a chemical analysis of that heap of manure and reproduce it synthetically, but in the exact chemical formula the result would not have the same stimulating and, above all else, the same healthy effect upon plant growth and fertility as the heap of manure itself. Most important of all, the synthetically produced fertilizer would contain no organic material, no potential humus without which all plant life in the end turns sickly or deficient.

I know that there will be those who say that all this may be disproved by the men who have grown certain plants in jars filled with chemicals in solution but I would still believe that the chemical men are wrong and that a race of men fed upon food raised in such a fashion would deteriorate for want of the mysterious elements which we do not yet understand. I know from my own experience that there is a vast difference in the flavor of a tomato grown in chemicals from one grown in warm, rich soil where the mysterious natural processes of the soil are uninhibited.

For me there is no more fascinating experience than the walks across field after field of the whole of Malabar, for each field tells a story

and never once have I walked over the land without having made a new discovery or learned some new fact.

On the ruined Bailey place, the latest of our acquisitions, three or four old square fields were thrown together for the purpose of laying out strips along the contours of the rolling glacial mounds. Today the whole process of restoring that land is still incomplete and the whole of the area in its present state provides an education in agriculture.

Much of the farm and the fields in question lie just below a high hill and from its summit you can look down and read the history of the place. Once it was a pretty good farm, perhaps as good as the Anson place, but by the time we got possession of it, it was about as ruined as a farm can be in this glacial Ohio country. For ten years it had been farmed by a man who believed, quite sincerely, that a farm was something to be mined, and that when the fertility gave out you went on and took another farm or preferably free government virgin land and in turn mined the fertility from that one.

This was the agricultural philosophy in which he had grown up, like the great number of American farmers still imbued with the pioneer tradition. That he made each year a little less money and never rose very much above the subsistence level did not seem to trouble him, nor did the fact that at the end of ten years he had wrecked the farm and sold it for many thousands of dollars less than he had paid for it, thus destroying his own capital. For his increasing poverty, he blamed the weather, labor trouble, bad luck and a dozen other things. But the reasons are very clear; the record of them is imprinted upon the fields spread out below the high hill, as clearly as if the whole story were written out. We have not yet succeeded in wiping out the whole of the record in order to leave our own in its place.

For a space of three hundred yards about the unpainted barn the grass and the crops are green and healthy. That was in the area over which during a period of ten years he distributed his barnyard manure. He never hauled it out of the barns until his cattle could no longer get in, and then he dumped it in the barnyard where it remained, leaching away its richness, until the cattle became mired and he was forced to haul it into the nearest fields where it has left the mark of its richness.

Beyond that circle the earth lay hard and packed, when we took over the place, with every knoll yellow and bare where the furrows have carried off water and precious topsoil. Here and there a furrow had grown into a gully. There was no humus, no decaying organic material to feed the worms, the bacteria, and to release slowly its supplies of nitrogen. It was dead soil.

In ten years, no crop of green manure was plowed under and no barnyard manure scattered over the fields. The man who farmed them believed that chemical fertilizer could make up for these things, that it was a short cut that could make farming easy and profitable. For ten years he used only chemical fertilizer. At first it was effective and the earth responded, but with each successive year the response grew less. It was like the effect of morphine upon an addict of many years. The crops began to look sickly and yellow and were subject both to disease and the attacks of insects. Meanwhile he left the soil given over to row crops bare all winter and the calcium, the phosphorus, the potassium leached out under the freezing, the thawing, the violent rains, until at last very nearly all that was left was the clay, the sand, the gravel which made up the basis of the soil. It had been reduced to something which might be described as a kind of cement, with traces of acid, virtually devoid of humus, of bacteria, of worms, of animal life of all kinds. Pheasants released on that farm fled it immediately for the richer fields near by. The quail abandoned it and at times it was difficult even to scare up a rabbit in the reaches of its barren desolation.

We have had the farm for a little more than two years and the problem of wartime labor has held us back from doing all we had planned to do with it. We have done what we could, which means that we cleaned the Augean stables, the barn and barnyard, and spread the manure as far as it would go over the fields. We have hauled and spread some lime, dehydrated lime in order to get quick results. We have run four or five strips following the contours across the field and these are planted in wheat.

The whole story of restoration is written there on the lower fields of the Bailey place. Where the powdered lime was first scattered the mixture of clover and alfalfa is thick and green. Where no lime has been spread the hay grows thin and mangy and wretched. Where manure has been spread across the wheat or where the sweet clover was plowed under, it is green and lush and vigorous. At one end, one of the long strips runs over into a field on the Fleming place where we had a good planting of alfalfa, part of which was plowed under to continue the strip, and where the alfalfa was plowed under, the wheat grows as green and vigorous as where barnyard manure has been spread. In the low spots between the knolls where the topsoil has been washed down and deposited, the wheat and the crops grow healthy and strong.

One day not too far off we shall have these fields looking uniformly green and fertile. There will be good crops on them and fat healthy cattle and the alfalfa and the corn grown there will fill mows and silos against the long winter and feed the beef cattle and dairy cows and chickens out of which comes the income for most of us at Malabar. Meanwhile, each time I cross these fields there is a new discovery and the new excitement of seeing land coming back, out of the degradation of bad farming into the realm of husbandry.

I had a Scottish grandfather of whom I have written much in *The Farm*. He ran away from home at sixteen to go to the California Gold Coast and never saw the inside of a schoolroom from that day onward. On a hundred acres of land, he raised and educated eight children and took care of many indigent relatives. He was one of the founders of the grange and one of the leading fighters in the struggle for more and better educational opportunities for children in a world not very far removed from the frontier. At ninety, he died one of the best educated men I have known, although he had left school at sixteen.

He was a specialist in agriculture and to be a specialist in agriculture he had to know many other things and perhaps best of all, his own land. There was scarcely a day passed in his life when his footsteps did not trace a pattern over his fields and orchards and woods, scarcely a day when he did not try some new if only small experiment. When, after he had passed the age of eighty and was crippled by a fall from his own great haymows, he managed to cover a large part of the farm in a wheel chair, pushed by one of his grandsons. He died while I was in France during the first World War but for me he has never died. Not only is he in my flesh and bones, he is in whatever spirit I possess. He goes with me over the fields of Malabar, watching them turn green and productive again, urging me to go on knowing the earth I possess more and more intimately. In a way, he lives behind all we have accomplished, all the trials we make. His spirit has gone on in my youngest daughter who has the same passionate love of fields and cattle and forest and stream.

I know now exactly how he must have felt when the experiments we make on our land with alfalfa succeeded, when each year the pastures carry more and more cattle, when the crops grow green across what were once gullies or bare spots where even weeds could not thrive.

Experimentation is the very essence of the "live" farmer. It is the enemy of stagnation and of evil practices, of the whole school of thought which believes that "if it was good enough for my grandfather it is good enough for me." Sometimes "Grandfather" was right in his methods, more often in Europe than in this country. That is why the farmer of recent European ancestry is, on the whole, so much better a husbandman than the American farmer of old native American stock. The European farmer is a good farmer because he is part traditionalist and part experimentalist. He is looking always for a better way, and his practices are far more in accord with the processes of Nature than the traditional land mining methods of the majority of American farmers.

At Malabar we have learned many things, some of them small and subtle, peculiar to our own soil and affecting us alone, some of them fundamental and important, not only to ourselves but to others. Some discoveries were the result of deliberate experiments. More often they were made by accident and the result simply of an eye which observed what went on under our feet and all about us.

There was the summer three years ago when a wet season prevented the final cultivation of the corn. The corn was well grown but far from free of weeds. The important fact was that it was above and ahead of the weeds. At first the weeds distressed us but as the hot, dry weeks of August and early September came on, I began to discover a remarkable thing—that where there were weeds in the corn rows, the soil beneath was moist and cool; where the corn was free of weeds, the soil was hot and parched and dry. Where there were weeds, the corn grew stronger and higher, the ears were larger and the grains weighed more. Much the same thing happened the following year with almost identical results. The discovery set under way a whole train of thinking that led us to new discoveries and opened up new fields of information always available, but not yet discovered by us.

The discovery and the observation were contrary to traditional and accepted beliefs—that weeds took moisture from the growing corn and that weeds of any sort or size impaired growth. Clearly in the case of our cornfields something had gone wrong, not with Nature but with tradition.

That weeds absorbed water and transferred moisture in the air we could not deny nor did we doubt it. Therefore it was necessary to discover how and why the soil beneath the weeds was still moist despite the added drain upon the moisture and why in the parts of the field that were free of weeds the soil was dry, dusty or caked. Shade, obviously, was one answer; the soil where weeds were mixed with the corn was more protected than those parts of the field where the corn alone provided a shifting, variegated shade and permitted evaporation through the force of the sun and wind. But further observation made evident another discovery which proved to be a far greater element in the fact of more moisture and better growth in the weedy portions of the field. We discovered that after a rain or a heavy thundershower, the water ran off the cleanly cultivated portions of the field and accumulated in the low spots. The corn in the low spots had too much water; that on the higher spots had little or none at all because it ran off. In the weedy corn, in the same field, on the same sort of mildly rolling land, no such thing occurred. No water accumulated in the low spots, no water ran off. It stayed where it fell.

At the Zanesville, Ohio, Soil Conservation Service station we found a part of the answer to the story of the moisture. It was simple enough, no more than this—that most of the rain which fell during the hot, dry weeks of late summer came with violence in the form of thunderstorms and cloudbursts, in drops which fell heavily upon the bare soil. Each drop, by its size and the impact upon the surface of the ground, sealed the surface of the earth and made it virtually impermeable to the heavy rainfall that followed. As a result, in the portions of the field where the corn was free of weeds the rain ran off to accumulate in the low spots where it half drowned the corn. Often enough on the higher ground a new rain would penetrate less than an inch.

In the parts of the field where the ground beneath the corn was weedy, the heavy drops of water never reached the soil, as great globules of water. Striking the corn and the leaves and stems of the weeds, each heavy drop was shattered and the rain reached the earth below, not in great dashing drops, but as a gentle, misty rain, like the rain to which northern and central Europe is accustomed. The parched, water-hungry soil was not sealed by the impact of the heavy rain; it drank up all the moisture as it fell, as a sponge absorbs water. There was no runoff water; it all stayed where it fell, evenly distributed over the whole of the weedy area. The tests on Zanesville experimental plots showed that where heavy rains fell on hot bare soil in midsummer as much as 70 per cent of the water ran off, seeking a lower level. On the plot next to it, covered by vegetation, 100 per cent of the rain remained where it fell.

Now the serious farmer need not deduct from the above that we advocate the cultivation of weeds with the corn; he may deduct if he likes that we at Malabar have no great objection to weeds in the corn provided the corn is kept clean until it is above and ahead of the weeds. And he may deduct also that we have no objection, after the corn picker has passed over the fields, to having in addition to the corn fodder a heavy amount of organic material left by the weeds to plow or disk into the soil for its lasting benefit and the benefit of the crop that follows. Many a good farmer sows rye grass between the rows of his corn for exactly the same purpose.

The discovery about the corn led on into further speculations beyond the field merely of rainfall and of moisture. For one thing, it led to speculation upon weeds and what is the nature of weeds. They have been variously defined, perhaps best as plants which grow where they are not wanted. Big purple violets in our flower garden are perhaps the worst weed we have to combat, yet in other parts of the country people exert great effort in order to raise the same sort of violet. Milkweed, once regarded as one of the most pestiferous of weeds, shows signs of

becoming one of our most valuable chemurgic-industrial crops. Its seed produces oil, its down as good a filler for life preservers and pillows and mattresses as the East Indian kapok. Its stem contains both latex from which rubber is made and fiber which can replace hemp and other fibers used in burlap and rope. The common ragweed, much maligned as a cause of hay fever, is with us a welcome visitor in a clover seeding. It serves to shade the young tender plants and it seems to have the same effect in breaking up heavy rainfall and distributing moisture as the weeds in the cornfield. It is even possible that there is some affinity with the young clover by which both plants benefit or that the ragweed has some capacity for improving the soil itself, especially poor soil. At any rate, I have no doubts about the benefit of ragweed to a seeding of young clover or alfalfa.

All of this led us then into the whole question of the affinities between plants. Nature never was a single-cropper and if a field is allowed to return to a state of Nature, certain plants will usually be found growing together. I know of no better example of this than the natural white clover and bluegrass rotation so common in our country. If a field of even moderate fertility is allowed to return to Nature in the form of pasture—that is if one ceases to cultivate it and turns the cattle in—white clover and bluegrass will make their appearance, perhaps together at first and then as the nitrogen in the soil becomes slowly exhausted, the bluegrass will begin to wane and at last virtually to disappear. Then the native wild white clover will take over and for two or three years it will reign until through the tiny nodules in the roots it will restore the nitrogen. Next the bluegrass begins to reappear and choke out the clover. For two or three years the pasture will be largely bluegrass until the nitrogen again becomes exhausted and the white clover moves in again.

No seeding is ever needed. Nature herself takes care of that in ways which to me still seem mysterious, although I am certain that with enough research there would be no mystery in the process. Occasional applications of lime and phosphorus serve to accentuate and strengthen the rotation and the quality of the pasture, but does not alter the fundamental affinity or process. We have many acres of bluegrass-white clover pastures which are as green and as weedless as the finest English lawn. So strong is the bluegrass-white clover affinity that no weed less tough than ironweed or thistle ever gains a foothold. Indeed the strength of the affinity at Malabar is so great that it has very nearly prevented us from growing the valuable Ladino clover as a crop.

Ladino is a super-giant cousin of the common wild white clover of the natural rotation and it restores nitrogen to the soil many times more rapidly than its miniature cousin, with the result that within two or three years the fields of Ladino which we planned to cut for hay or protein supplement or pasture, are largely taken over by the eager bluegrass.

Certainly there are other affinities in the crop world. One of them is unquestionably in our observation that between soybeans and corn. Again the observation of this fact came about by the accident of the two crops becoming mixed where two fields overlapped. Where they happened to grow together on the overlapping strip both soybeans and corn were sturdier and healthier than in the immediately adjoining fields where they grew separately as single crops. The next year we drilled soybeans in the same rows with corn and cut them together to put in the silo and made the finest of balanced silages for both dairy cows and beef cattle. The results were the same as in the accidental experiment; both crops were sturdier and healthier than when grown separately. We have gone further and drilled soybeans not only in the rows but between the rows of corn at the final cultivation. The results are good and fairly conclusive and the resulting ensilage can scarcely be improved upon for beef cattle. The hay beans grow three to four feet in height, even producing small atavistic tendrils, indicating that once long ago the soybean was a climbing vine and the production of seed is considerably higher than that of hay beans grown separately as a single crop

The reasons, I think, are fairly obvious. The beans planted were tall hay beans, not those grown for grain. The hay beans have roots that thrust deep, sometimes as deep as thirty inches, especially when grown with corn, and being a legume the roots are covered with scores of little nitrogen-fixing nodules. As the hay beans are cut green before the beans have ripened, there is little of the drain upon minerals in the soil which occurs during the ripening process. We have thus used soybeans to replace the ordinary field weeds as a soil cover. The hay beans not only put nitrogen in large quantities into the soil but serve to shade it, keeping it cool and moist and to shatter the great, heavy raindrops of July and August and reduce them to a mist which the thirsty soil can absorb. The reasons why the hay beans benefit the corn are easy enough to discover but why the corn benefits the beans, I do not know, unless it is that they like the cool of partial shade during the hot months. All this comes within the realm of the affinities between plants about which too little is known and concerning which too little research has been made. I only know the fact of which my eyes are the witness. We have achieved a better balanced ensilage, grown a greater tonnage upon the same acreage, and benefitted the soil by an increase in nitrogen, conserved rainfall, and checked row-crop erosion, all by the utilization of an affinity between plants.

We have begun experimental plantings of corn with vetch and cowpeas, both legumes of which the seed is less expensive than that of hay beans, and are trying buckwheat, which is not a legume, but serves to feed the bees and makes excellent ensilage with corn. The results to date are encouraging, but not definitive.

The affinity between corn and the legume family is nothing especially new. In the South the poor hill farmers rarely plant corn without cowpeas, and long before them the Indians knew the fact of the affinity between legumes and corn although they did not understand that it was the nitrogen which made the corn sturdy and green. They rarely planted corn without planting beans in the same hill.

Scarcely a day passes without some new observation in the so-called miraculous but really only simple operations of the laws of Nature. Some of them appear throughout this book and many in the chapter that immediately follows this one. That is so because the chapter concerns Edward Faulkner and he too has been a man who read his small plot of earth. Neither in agriculture, nor in economics, nor in human history does anyone ever succeed in creating a plan and then *imposing* a plan or theory. In all the progress of mankind, in all his discoveries, one thing grows out of another in perfect continuity, based upon observation and the utilization of what has been observed, always within the limits and benefits of natural law.

e view from *Mount Jeez,* Malabar Farm

"My ninety acres"

I have a friend, a little old man who lives on the hill in Possum Run Valley in a small white house on a farm which is known as "~~My~~ Ninety Acres." It has never been named that as farms are named "Sunnyside" or "Shady Grove". ~~It has no sign bearing that~~ The name is not painted on a red barn nor on a fancy sign hanging ~~by~~ at the end of the lane leading up to the house, but throughout the valley everybody always refers to Walter ~~Oat Oates~~ Oakes farm as "My ninety acres." At first, years ago

selection from the original manuscript of *Pleasant Valley*

The Big House

All photographs courtesy of the State of Ohio: Department of Natural Resources, The Ohio State University & the Bromfield family

Bromfield & holstein in front of the Big House

romfield family at home in *Pleasant Valley*

Louis Bromfield working on terraced gardens

pring-fed pond & "swimming hole"

outbuildings

"The roadside market to end all roadside markets"

Louis Bromfield at his spring-cooled vegetable stand (insert: detail of trough)

"The farmer is, I think, the happiest of men for he inhabits a world that is full of wonder and excitement"

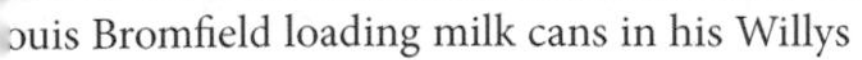

ouis Bromfield loading milk cans in his Willys

ıain cattle barn at Malabar Farm (since rebuilt as a timber-frame structure)

Louis Bromfield inspiring a younger generation in the virtues of farming. Insert: Bromfield milking a guernsey

Louis Bromfield on farm tractor "Oh, What a Beautiful Mornin'"

ouis Bromfield's "working" study

aughter Ellen shelving
reserves

Bauer–Toland Studios, Detroit, Michigan

George Hawkins, Bromfield & Prince • Portrait by J. Anthony Wills

The Bromfield family sitting down to a meal in the Big House diningroom with Mary and the Bromfield children on the right and Bromfield's parents on the left. If this photo is from 1940 or 1941, that is probably a young Chester Hines serving. Hines left the Bromfield household for Los Angeles and a distinguished writing career of his own.

This is the William Ferguson homestead which burned down before Bromfield acquired the property. The porch steps, now in deep woods, are a reminder of the Newville Road which once ran in front of the house.

The Fleming place, the working farm of today, as seen in 1943. The Sweitzer-style barn was built in 1837 about the same time as Shrack built his saw- and grist mill across the road. The barn has both scribe rule and square rule construction.

Louis Bromfield,
ulitzer Prize
ward-winning
riter teams up
ith Ernest
lemingway to
ring to the screen
eally fine writing
nd story-telling.
romfield prefers
o work at his
ooo acre farm
ecause, he
ays—a man can
rite better in the
ountry."

Malcolm Bulloch, Paramount Photo

Eugene Robert Fri[illegible]e, Warner Bros.

Humphrey Bogart & Lauren Bacall—frequent guests at Malabar Farm

THE BUSINESS OF *PLOWMAN'S FOLLY*

ONE MORNING NOT LONG AFTER THE FARM WAS STARTED there came to the farm office a smallish, graying man with very bright blue eyes. He said his name was Faulkner and that he wanted to talk to me about a new theory of cultivation which did away altogether with the conventional, long accepted moldboard plow. Because I am articulate and because I frequently write on subjects other than fiction, a great many messiahs have come my way, so many that there are times when I am inclined to believe there are more unbalanced people and more cranks in the United States than in any other country in the world. No nation has ever produced so many economic panaceas, so many plans for doing away with work and getting something for nothing. During the long course of the New Deal more than one reformer associated with it has brought forward fantasies of thought and planning no more related to reality or fact than the ideas of the Townsendites, the Dowieites, the House of David or any of the other countless economic and religious groups which have found the answer to everything. These messiahs call upon me and write me long and occasionally abusive letters and so, slowly over a period of years, I have developed a kind of phobia about them. In their presence, fixed by the half-mad look in their glittering eyes, I feel my heartbeat slow down. My face turns gray and presently I am seized with indigestion. All of this, no doubt, is no more than hysteria born of boredom, because for me nothing is so boring as a crank with a single-track mind, especially when his plan is to take the human race by the scruff of the neck and force it to follow out his own half-mad ideas.

And so when the man who said his name was Faulkner and that he was for abolishing the plow, came into my office, the old symptoms began to make themselves evident. I listened with only half my mind and said "yes" and "very interesting, I'm sure" and at last bade him good-by and got back to my own work, thinking I should never see him again.

But Ed Faulkner was a persistent fellow and rightly so, for he had discovered, for himself and entirely by himself, some important and fundamental truths and he was not prepared to have them overlooked. In short, Ed Faulkner came back, and when he came back I began to listen and as I listened, I began to be interested and to understand why it was that he had had a hard time getting a hearing.

His ideas were in one sense revolutionary; in another sense they were as old as time and as old as Nature herself. The revolutionary ones were not likely to be welcomed by either bureau men or by the academic world of agriculture. They were upsetting because they ran counter to much that farmers had been taught by bureau men and professors over a long period of time. Someday, if there is ever time, I should like to compile a whole book of the wrong and destructive things which have been taught in American agriculture, with a second volume dedicated to the things which have been taught more for the benefit of manufacturers of farm machinery, of chemical fertilizers and of prepared and expensive feeds than for the good of the earth or the welfare of the farmer. Not all the false things were taught deliberately by men who knew they were false; some were mistaken, others merely stupid, a few were prejudiced cranks and a good many spent their academic years simply in looking for the short cuts which exist neither in agriculture nor in Nature. Not all bureau men and professors are bigots—to many of them our civilization owes great debts; but the mere fact of sitting long behind desks lecturing young people or of passing years on a swivel chair far from the earth out of which all agricultural knowledge must come, does tend to produce an inflexible and sometimes a closed and prejudiced mind.

So when a man proposed to abolish the plow, it was not likely that he would meet with a sympathetic reception from circles which had acclaimed the plow for at least one hundred and fifty years as a great instrument of civilization and of man's welfare.

As Faulkner talked, my interest grew because what he proposed seemed so reasonable and because I knew out of my own experience, both in gardening on a small intensive scale and in farming on a large scale, that much of what he was saying was true, with that truth which lies in the very processes of Nature.

He had been a county agent and had resigned because of differences with the Extension Service authorities over his unorthodox ideas. From then on he had worked at various jobs, mostly at the insurance business, which he found not too profitable. In his own town of Elyria, Ohio, forty miles from Malabar, he was regarded by those who knew him as a man with an obsession. He was the son of a Kentucky farmer, whose own farm had remained green and fertile while the neighboring farms slipped farther and farther down the scale of production until many of them were abandoned. In a way, Faulkner had given all his life to the earth and what it could teach him. When I first saw him he was conducting his experiments with soil in his own back yard and on two acres of poor land which he rented by the year, a testing ground that was obviously inadequate in terms of commercial farming. Nevertheless, on that little piece of land he had arrived at proving many things with which I found myself in agreement, not the least of which was that given reasonably good subsoil, topsoil could be rapidly rebuilt, not by short-cutting Nature but by adopting her own methods and by speeding up and intensifying them. And Faulkner, like myself, was a great believer in the virtues of mulch as a means of growing crops and doing away with much labor and cultivation.

And so I listened and presently I too began to talk. I, at least, was learning something. I had never thought of the evil the moldboard plow might do until I listened to Ed Faulkner.

And then one day he brought me a thick manuscript in which he had set down all his ideas and the results of his experiments. It was a very long book and not too well organized, not nearly so well expressed as when Faulkner talked and the light came into his blue eyes. He wanted help and advice with it and I had to tell him that I should only be deceiving him if I told him that I could find the time to do a conscientious job on it. And so he left me, and Ollie Fink, who is in charge of conservation education in Ohio, and Paul Sears, head of the botany department at Oberlin College and author of *Deserts on the March*, gave him the needed professional help and advice. The manuscript was rejected by four or five of the largest publishing houses, all on what I must say were reasonable grounds, commercially speaking—that few people, even farmers, would be interested in a book on plowing. I heard no more of the book or of Faulkner until one day I received a thinnish nicely printed book called *Plowman's Folly* by Edward H. Faulkner. The Oklahoma University Press and its active manager, Savoie Lottinville, prompted by Paul Sears, had had the courage to publish it. I stayed up that night until I had finished the book. I knew that Faulkner had something important to say but still doubted that it would receive attention.

I was wrong. Within a few weeks, half the publications of the country had published editorials concerning it, even such unlikely ones as the *New Yorker* magazine and the New York *Daily News*, which devoted a full column editorial to the subject. Everywhere you went you heard people talking about it, from practical, solid farmers to two Hollywood actresses who in a Chicago hotel asked me "What is all this business about *Plowman's Folly*?" I myself did a piece about Faulkner for a national magazine and was forced to set up a special post-card department to answer queries as to where the book could be purchased. The queries came from every part of the country, from servicemen abroad and from farmers in South Africa, in England, in Persia, in Canada, Australia, South Africa and Palestine. The presses could not turn out copies fast enough for the demand and the war restriction on paper presently made it impossible for the Oklahoma University Press to meet the demand. Another publishing house then took over the book and the latest edition scheduled is for 250,000 copies, far more than the sale of most best-selling novels. Mr. Faulkner was engaged as agricultural consultant by one of the great radio corporations and today receives more invitations to lecture than he is able to fulfill. In a recent letter I received from him, he wrote "I have been forced to increase my lecture fee in order to control the demand."

The story of Ed Faulkner is the story of a man with an idea, who stuck to it. It is essentially a success story, almost a fairy tale in its happy ending. The reasons for the success of the book are many. It came out at a time when there was more interest in agriculture than ever before in the history of the nation. It came also at the moment when the agricultural revolution, which has been going on slowly and imperceptibly in America for more than a quarter of a century, was beginning to take form as a fact. It came at a moment when there was an immense and rising wave of popular concern over the conservation of our natural resources, and at a moment when many intelligent farmers were looking for something that they were not getting, either out of the Department of Agriculture or the agricultural colleges, something which had to do with what is perhaps the fundamental fact of our national existence, the soil. I think that is also the reason why there has been so great and widespread a demand for Faulkner to talk to farmer groups.

Not all his readers have been practical dirt farmers; many of them have come from women's garden clubs, and it would be foolish to underestimate the knowledge of garden club members regarding the soil and even agriculture in general. Many of them know far more about soil than most second-rate farmers. And many of Faulkner's readers have come from among the so-called city farmers whose knowledge and initiative and brains it would also be unwise to underestimate. Most of them are successful in their own fields because they are men of brains, ideas and initiative, and many of them at middle age or later have transferred their brains, energy and initiative from business and professions to the realm of stock-breeding and agriculture.

The city farmer is not content with farming as his grandfather farmed. He does not consider it good enough. City farmers are looking for new and sound ideas in agriculture. They are making experiments of great value which the average dirt farmer is unwilling to make either because the experiments represent financial risks which they dare not take or because they lack the initiative or the energy to make them. It was only natural that *Plowman's Folly* should find many readers in their ranks.

At Malabar we do not agree with all of Faulkner's original theories and since the publication of *Plowman's Folly* he has himself modified some of them. I do believe that he is fundamentally right, that the moldboard plow has wreaked great damage and that it is still doing so. More than that, the type of agriculture and cultivation it induces is largely an unnatural and faulty one. In one sense, the moldboard plow is a short-cut implement. It was designed to do too quickly and too neatly what cannot be done too quickly and too neatly without grave damage in the long run.

Speaking broadly, Mr. Faulkner's case against the plow is that it turns over and buries all protective mulch on the surface of the earth and leaves the soil exposed and bare to the burning sun and invites destructive erosion both by wind and water. More than that he contends that by turning over the surface trash, whether it be weeds, sod, green or animal manure, the moldboard plow packs the surface trash into a narrow layer subject to great pressure both from the weight of the soil above and the weight of the tractor and machinery passing over it. This pressure produces heat and fermentation rather than decay (which is the natural and beneficial process), creates harmful acids and reduces the production and availability of beneficial nitrogen. Still further the compacted layer of trash serves to create a barrier which prevents moisture from rising from the subsoil below to the roots of the plants growing above and prevents the roots of the plants from seeking the moisture that lies below. At the same time the earth above the compacted layer is left bare to all the drying process by wind and sun, the erosive processes of wind and water, thus creating a condition of artificial drought between the surface of the soil and the compacted layer of organic material below.

With all of that, I think any thinking farmer, indeed anyone familiar with the mysteries of the soil, will agree. I do not believe, however, that the plow can be done away with entirely. On our own farm where in the garden and the fields we have largely practiced *tillage* as well as plowing, there are cases where the old-fashioned plow is the only implement which can do the job. On our poorer fields on the Bailey place where bad farming reduced some of the fields to a kind of cement, only an old-fashioned plow can break up the soil and permit the mixing of organic material to improve it. And there are cases where sod is so heavy and tough that it is not possible in terms of labor and time to dispense with the plow. It would require too many days of disking to fit the field for planting. And there are many conditions of soil and of climate which render tillage rather than plowing impractical in commercial farming at least with any tillage implements now in existence. Nevertheless Mr. Faulkner is right. The plow has done great damage. While we at Malabar are not implacable enemies of the plow, we are enemies of *good plowing*; the old-fashioned kind which buries everything and leaves the earth bare as a bone.

The single-crop wheat country was the first to discover that the use of the moldboard plow was rapidly destroying productivity and the soil itself. In the first fortunate years when the soil of the wheat country was deep and virgin and apparently inexhaustible, it was the habit to burn the straw to get it out of the way and then plow the fields, fit them and plant. This process was the perfect example of taking everything from the soil in a single-crop agriculture and returning nothing to it. Nowhere in history, save perhaps in the single-crop cotton and tobacco areas, has mankind made such a record for the destruction of soil as during the first forty or fifty years of wheat farming in this country. Largely speaking, the great wheat area is one of little rainfall, and after a number of years this process of mining the soil brought about a rapid loss of humus, with a consequent rapid decline of production. More than that, with little or no humus, the soil dried out and blew away and the desert storms began in the wheat area just as they came to China for similar reasons two thousand years earlier. In the thirties, billions of tons of soil from the wheat country were blown eastward to be deposited over cities and farms as far east as the Atlantic Ocean. Obviously, unless something was done, the whole of the great wheat producing belt was destined to become a desert like the Gobi and large areas of China. And the first step was to do away with the moldboard plow which turned over all protective straw, weeds and trash, and left the soil bare to wind and sun.

Today the soil of nearly all the wheat country is prepared, not by plowing but by the use of heavy disks or disk plows which do not turn over the earth and bury the trash but rather chop it up into the soil, leaving a mulch on the surface or chopped *into* the surface which prevents not only the evaporation of the precious rainfall of the wheat country but checks or prevents the blowing or washing away of the soil. Soil Conservation Department experts estimate that the velocity of a fifty-mile wind can be reduced to something under ten miles per hour at the surface of the earth by the presence in the soil of thin, chopped-up mulch. And in the wheat country, good farmers no longer burn the straw either in the fields or in the stacks, but chop it back into the soil.

The efficiency of this non-plowing surface trash method of preparing the soil in relation to erosion by water has been proved in our own fields many times. In our Valley all land is rolling or hilly country and some of our fields cannot be properly contoured or farmed in strips because the hills and slopes are neither wide nor long. They slant in many directions. Nevertheless these fields must produce crops if we are to farm at all and that is where trash-farming has proved a salvation. Our roughest, steepest terrain is kept in permanent bluegrass, white clover pastures. The next steepest fields—the ones *possible* for cultivation are utilized to grow rotations of alfalfa hay and small grains, in our case, wheat and oats. Only on land which can be strip-farmed or contoured do we grow row crops or soybeans. The middle category of land is protected by hay crops from all erosion but periodically we find it necessary to put in a crop of wheat or oats in order to get a new seeding of hay. In theory the small grains, wheat or oats or barley or rye, protect the soil from erosion in winter or during heavy rains, but we have found that the theory by no means proves itself absolutely, especially during the long winters or before the small grains have grown sufficiently to cover the soil and break up the impact of heavy raindrops. I am, of course, speaking of fields which have been plowed before seeding to small grains, leaving them bare and without any protection by mulch or chopped-in trash. Always on such fields we have had losses of both rainfall and soil and the formation of small gullies.

Only when we ceased plowing these fields did we find the absolute check to loss of rainfall and topsoil. The preparation of a field that has been in hay sod by disking is a long and tedious process, sometimes not economically possible from the point of view of time and cost in labor and gasoline and tires. A disk plow, well managed will do the job of fitting the field without leaving the soil bare, but not so efficiently as we wished. The solution came with the use of a tiller, a kind of cultivator similar to those used in orchard cultivation. It enabled us to rip up the soil without turning it over and without burying the sod, roots and trash. Once or twice over with the disk after the field had been roughly torn up, permitted time for the sod to die, and left the field with a surface mulch of rotting roots and stems which acted all through the winter as a kind of sponge. *No* water runs off those steeply sloping fields. *No* topsoil is lost and no small gullies form. As a means of checking erosion either by wind or water, the method has, with us, proved absolute.

But there were other advantages, profitable especially upon a farm like ours which consumes large quantities of good hay. The seedings of clover and alfalfa on these fields showed a germination of very nearly

100 per cent as against the 50 per cent or 60 per cent which we were able to obtain on fields plowed and left bare save for the slowly growing winter wheat or even the more rapidly growing oats. This enabled us to cut the expense of clover or alfalfa seedlings by 40 per cent and still we had a thicker and more even stand of seedlings than we were able to obtain on fields that had been plowed. The results were the same whether the seeding was drilled or broadcast. The results have been very nearly the same whether there was plenty of rain or in a dry season.

An examination of the surface of the field told its own story. On the fields which had been trash-farmed the whole surface was covered by a mat of decaying vegetation mixed *with* the soil and this served to prevent evaporation and conserve moisture not only in the soil beneath but actually in the mulch itself. Wherever a tiny seed fell, there was moisture enough in the mulch to bring about germination. No seed fell on hard, bare ground dried out by the first day of sun and wind, for there was no hard, bare ground. No seeds were buried too deep or choked by clods of earth. No seeds were washed across the surface of the soil to collect all in one place in a low spot or between the little ridges left by the drilling of the wheat the autumn before. It was evenly distributed, it stayed where it fell, and the mulch always provided *natural* conditions for the germination of alfalfa and clover seed and Nature, in response, rewarded us richly.

It is true that in wet seasons or when we made a late and hurried fitting of the soil, not all the sod chopped and ripped into the earth died. We sometimes found timothy or brome grass or even clover and alfalfa growing up with the ripening wheat, making it difficult or impossible to combine the following season. We are not in wheat country and do not raise wheat as a crop save as a means of getting in a moderately profitable extra crop while reseeding a field to hay, and so this fact did not disturb us, especially since the mixture of grass and wheat, cut when the wheat was in the milk gave us a crop of excellent protein silage far more valuable to a dairy, sheep and beef cattle farm like ours than the wheat itself.

One charge made against trash-farming in the Faulkner way is that it produces an abnormal crop of weeds, and the charge is not without foundation although ground infested with weeds will be weedy whether plowed or trash-farmed. Seeds of the commonest weeds will lie dormant if buried deep enough by the plow and germinate a year later when the plowing process brings them again to the surface. After combining the wheat or oats on trash-farmed ground, there is occasionally a thick crop of weeds, which in a wet season threaten at times to choke out clover, timothy, brome grass and alfalfa seedlings. But we have managed to turn this crop of weeds from a menace into a benefit by the use of the mowing machine. One clipping, or at most two, cuts down the weeds before they have reached the seeding stage, and left on the field the clipped weeds provide a mulch for the seedlings, shading the earth from hot wind and sun, keeping it loose and moist to promote the growth of the seedlings before the arrival of winter. The same mulch by insulating the earth tends to keep the ground *frozen* throughout the winter and prevent the alternate freezing and thawing process so disastrous to clover and alfalfa seedlings in the more northern regions. As the clipping process continues year after year the fields actually become almost clear of weeds, a condition I am not altogether certain is good for them.

One discovery in agriculture often leads to another and very often men or women unknown to each other are engaged in making observations or experiments moving in the same general direction, and so we found that while we were making our own experiments with trash-farming and methods which did not employ the conventional plow, others were working in the same direction—Mr. Faulkner in his back yard and his leased two acres; Dr. H.L. Borst in the fields of the United States Soil Conservation Service at Zanesville, Ohio; Christopher Gallup on a small Connecticut farm; men at agricultural colleges in the great wheat factory of the far Middle West; and Mack Gowdy and a few intelligent "live" farmers in the South.

The experiments in trash-farming led us into discoveries regarding the best way to grow alfalfa in our part of America. In the past we had been taught that in order to grow alfalfa it was necessary to make a careful preparation of the ground, adding lime and phosphate gradually sometimes over a period as long as three or four years before seeding. Next we were taught that the way to plant it was to prepare a powder-fine seedbed and make the seeding in late August or September on ground that had been plowed and left bare.

We never had any success with this method, perhaps because the light soil of our glacial hills dried out too much during the hot months of late summer. I have rarely seen a really good crop of alfalfa grown by this method in rolling or hilly country, but only in bottom land where there is enough moisture to insure germination; but on such ground, the alfalfa, which insists on good drainage, often suffered later on by being drowned out when the rains of spring and winter came along. Also, having made only a feeble growth by the time our winter came along, the young plants were decimated by the process of freezing, thawing and "heaving" out of the ground.

This method of alfalfa culture was the traditional one in the dry and irrigated lands of the West where alfalfa first became a profitable commercial crop, where drainage was good and moisture could be controlled and where there was in most regions, little or no winter and the problem of "heaving" was nonexistent. Apparently, it had never occurred to the men to whom it should have occurred, that the culture of alfalfa might be a quite different problem in Ohio and in California. In any case, the farmers of Ohio went on struggling to grow alfalfa in the "California way," sometimes, when all the conditions of moisture and weather were right, with success—more often without, until many of them came to abandon alfalfa altogether; although, intelligently, they recognized its great value as the finest of protein hay with the possible exception of the virtually unknown Ladino.

At Malabar, we found the whole business disturbing. In the first place the whole approved process seemed artificial in a country where the rainfall was well distributed and other legumes, notably the clovers, flourished naturally along the roadside. I had a personal theory with which Max, with his agricultural college education, did not agree. It was a revolutionary one to be sure—simply that alfalfa, being virtually the nearest relative to the common sweet clover which flourished on the clay of any open roadside cut in our country, was a natural in Ohio if treated properly, and that it was really a "poor land" crop and a soil improver. I did not believe that all the fuss and bother was necessary to grow good crops of alfalfa. More than that, I suspected that all the pampering was unnatural and actually harmful.

At the same time Dr. Borst at Zanesville was working along the same lines, quietly, without announcing his results until he became sure of them. Without any direct scientific experimentation, but only by observation, we on our side began to arrive at the same results he was achieving.

We did it by watching the fields, the fence rows and the roadsides, where alfalfa grew as easily as our native sweet clover if it was simply let alone.

We discovered that a spring seeding, rather than one made in late summer or autumn, done simply as we made seedings of clover, but on trash-farmed fields rather than on the bare, plowed soil, achieved results. We knew that lime and phosphorus were considered essential and helped, but presently we made another startling discovery—that alfalfa would thrive on poor soil in our county even *without* applications of lime and phosphorus; in other words, that in our county it was definitely a poor soil crop. Poor soil to be sure is relative and the subsoil of our glacial hills all the way through is potentially good soil as compared to many subsoils elsewhere. Nevertheless, the field we chose for the "accidental experiment" was the worst field on the four farms. Owned impersonally through a defaulted mortgage by a bank in a distant town, it had been rented out to the first comer for nearly twenty years, with everything taken off it and nothing put back until even the neighbors no longer found the fifty acres worth farming at a rental price of five dollars a year.

Our discovery came about as an accident as so many discoveries in agriculture have come about. We set out merely to put the weed-grown field in order by plowing under in the spring the accumulated mess of goldenrod, wild carrot, other weeds and even sumac which had taken over. It was impossible to do a good job of plowing because of the accumulated trash, and the weedy sod stood on edge behind the plow rather than falling over to bury the trash. This troubled Kenneth who did the job because he was ashamed of not doing what in our neighborhood was accepted as good plowing, which meant that *everything* was buried and the earth left free and bare of all trash. As it turned out, the "bad plowing" which troubled Kenneth was a very "fortunate accident" and in connection with other accidents and experiments finally convinced us that the traditional clean "good plowing" of our neighborhood was, so far as we were concerned, the worst kind of plowing.

In order to whip the rough, weedy field into shape we disked it roughly with the spring tooth following the disk. The results were not good so far as "fitting" went. The seedbed was rough and trashy but we only wanted to get the field in order and as it was already late in the season, we hastily drilled in a seeding of oats. Immediately afterward, we broadcast a seeding of the legumes we had left in the seed bin. We hoped, not too optimistically, to get a cover crop of legumes which we could plow under as the first stage in the rehabilitation of the field. Fortunately, the contents of the seed bin was mostly alfalfa with a little red clover and a little mammoth clover, a little alsike and a little brome grass. In order to give the seeding the best opportunity the seed was inoculated with nitrogen-fixing bacteria. The oats were given in drilling the average amount of fertilizer, nitrogen, potassium and phosphorus in the hope that we should get a crop that would be worth combining.

All the conditions were as far apart as possible from the traditional instructions on seeding and growing alfalfa. One natural element was in our favor—that we had plenty of rainfall after the seeding and the percentage of germination was high, especially since the seeding was made while the ground was still open and the seeds worked their way into the soil with the first rain. But germination did not necessarily mean success; seed will germinate in damp cotton or in pure, moist sand. The real problem was whether the tiny seedlings could find in that poor soil the elements necessary to grow into sturdy, hay-producing, soil-improving plants.

The moisture helped them to a good start. The soil was so poor that the oats were not worth combining and we left them in the field, to ripen or die, eventually leaving a residue of broken and rotting oats straw above and among the alfalfa and clover seedlings. As the summer progressed, the seedlings, especially the red clover, made an astonishing growth, wholly inexplicable in view of the poorness of the soil and the lack of lime. No doubt the excellent moisture contributed to the growth and the trash left on the field by the "poor plowing" job conserved moisture and, in decaying, provided nitrogen until the seedlings had established their own nitrogen-fixing nodules. In any case, by autumn there was a rich growth of legumes over the whole field to the height of eight or nine inches. Mixed with it, there was a large amount of ragweed which led us again to suspect some sort of affinity between ragweed and clover which benefitted the clover, some beneficial influence beyond the mere fact that the ragweed provided a certain amount of shade and protection to the young seedlings during the hot weeks of August and early September.

Despite the agreeable outlook, we knew our problem of alfalfa as a poor-land crop had not yet been solved. There was still the winter to be passed, with its perpetual freezing and thawing, which could cut our prospective crop in half by heaving the plants out of the ground.

The winter of 1940 was one of the hardest I have ever known. Freezing and thawing alternated during most of the weeks of the whole season and by spring some of our best fields of clover or mixed hay were white with the roots which had been forced out of the ground by the variations of temperature. Oddly, the poor field, where our unconventional operations took place, suffered not at all. There was *no* heaving and our seeding came through as thick as it had been at the beginning of the winter.

Looking for the answer in the field itself, we arrived at a theory which during the three years since has proven infallible. It was this—that the decaying oat straw left in the field, the residue of ragweed stalks and the dead vegetation of the clover and alfalfa plants which were not clipped had provided a mat of mulch over the whole field, covering the roots of the plants and acting as an insulator to check the freeze-thaw process which had been so disastrous in other *well plowed* bare fields with much better soil. In other words, the earth itself had become frozen early in December and it *remained* frozen throughout the bad winter because the mulch of dead and decaying vegetation prevented any warm spell or burst of sunlight from reaching the soil and thawing it.

The field by hay-cutting time revealed a thick, vigorous stand of red and mammoth clover and alfalfa and we cut some of our finest hay from it, although we were a little disappointed that there appeared to be much more clover than alfalfa. There was no ragweed whatever. At the second cutting, however, the alfalfa showed up. There was a little clover and about the right amount of brome grass, and a thick stand of alfalfa, not 100 per cent, but an extraordinary stand considering the fact that we had violated all the rules. The thick growth we did not cut for hay, but pastured lightly, permitting the beef cattle to cover the field fairly thoroughly with droppings and urine. The field went into the winter with the decaying mat of vegetation still in evidence between the alfalfa plants.

The following winter was less difficult than the preceding one and again the field came through with no loss from heaving. In the second year the alfalfa plants had increased in size sending out ten or twelve stems of new growth where in the field year there had been only six or seven stems. The red clover was nearly choked out. The brome grass remained and here again there appeared to be some affinity between the brome grass and the alfalfa. Apparently they liked living together and each contributed something to the health and vigor of the other. Despite the loss of the clover from the field, the yield from the thickening brome grass and alfalfa gave us more hay than in the preceding year.

Then came a remarkable discovery. In walking over the field in late May of the following year, Bob and I discovered something neither of us had ever seen before in a field of mixed hay. On the by now virtually decayed surface mulch between the alfalfa plants there had appeared millions of tiny seedlings of red clover and alfalfa. As the summer progressed, they continued to grow and we made our first cutting of hay early in order to let in the sun and encourage the growth of the volunteer seedlings. Wherever the alfalfa had developed into a 100 per cent stand the seedlings were choked out but where the stand was thinnish the seedlings developed into sturdy plants and actually *thickened* the stand of mixed hay in the third year by the production of young new plants.

Whence these new seedlings in an established stand of alfalfa came from I do not know. I only know that they appeared, that some grew into mature plants and increased the stand. Either they were seeds from the original sowing which remained dormant for a period of time or they were seeds from the alfalfa plants already established which seems unlikely in view of the history of the field, with two hay cuttings and a light pasturing in the late summer season. Also any theory that they were distributed by birds is unlikely in view of the perfectly even distribution of the new seedlings over the entire field. I do know that I have never observed the appearance of seedlings in such extraordinary numbers in fields of alfalfa where the ground was prepared in the conventional fashion and left bare to the elements. It seems true beyond much doubt that the mulch of oat straw, dead ragweed and unclipped clover growth, provided a damp layer of decaying vegetation which encouraged germination and growth, by keeping the soil below moist and cool in summer and immune to the evil results of freezing and thawing in winter.

The observations led to other speculations for which we have been unable to find definitive answers. Had the poor land ragweed, for example, any effect upon the vigorous growth of the original seeding and ultimate germination of dormant seeds? Could it be that ragweed in itself possessed soil-restoring qualities or some special affinity for legumes? During the first seedling season, it served to shade and protect the seedlings and vanished utterly during the second season. Does ragweed, which flourishes in poor soil and on bare fields, have a place in the natural scheme of things for the improvement and restoration of rundown land? Does the ragweed produce any vitamin, hormone or acid which stimulates germination? Did the distribution of animal manure and urine over the whole field during a period of pasturing provide hormones, vitamins or other organic animal gland secretions, which stimulated not only growth of the adult plants but the germination of dormant and reluctant seeds?

Certain things we know—that the experiment had the benefit of good rain distribution and that the *apparent* absence of lime and phosphorus from the thin, leached-out topsoil did not prevent the establishment of a fine stand of alfalfa on poor land. It is highly probable that the favorable growing season and the thinness or absence of any real topsoil permitted the roots to thrust their way quickly into the subsoil of our glacial hills where the plants found, in the great mountains of glacial soil scraped across half the American continent, not only lime and phosphorus but all the other elements, down to trace minerals, which they needed for growth and health. This speculation is both interesting and important in view of the fact that the stand of alfalfa has always been remarkably healthy and free of any of the diseases which can attack alfalfa subject to malnutrition or lack of a balanced soil diet. The fact remains that a wretched field produced for us an excellent stand of alfalfa and brome grass, from which we have made three cuttings a year of excellent hay for three years, without any sign of weakening the stand.

All of these speculations are largely, of course, in the realm of things not yet discovered or determined by scientists and may well lead to new and important ideas concerning agriculture. A few men are already working on them, confronted by an opposition based upon complacency, conservatism and in some cases, ignorance.

So far as we at Malabar were concerned, the rather carelessly initiated "accident" proved two things—that so far as our Ohio land was concerned, the whole traditional culture of alfalfa was wrong and wrong, fundamentally because the whole process—the powder-thin seedbed, the late summer sowing, the exposure of the immature seedlings to the damage of freezing and thawing through the long Ohio winters, the growing of alfalfa as a single crop (without the ragweed, the brome grass and even the red clover) were unnatural and in actual contradiction to the natural habits of alfalfa. What struck us as important was that the whole of the process we used was a natural one—the seeding, the ragweed, the rough plowing and careless fitting, the natural protective mulch, even the pasturing part of the process. Save that we have substituted tiller and disk for plow in fitting the fields, we have used the same process since then in establishing alfalfa and the results have been uniformly excellent. The only changes have been that we have added lime and occasionally phosphorus, although I am not yet convinced that with the quantity and quality of the glacial subsoil peculiar to our region that this was absolutely necessary. It has very likely hastened the growth and stimulated the health of the alfalfa.

It was a part of this field which by necessity became incorporated in wheatland when new strips and contours were laid out. The wheat sown in the ground which had been a part of this field grew rich and dark green and yielded three times as much wheat as the ground adjoining it. The line between the good wheat on the plowed-up alfalfa ground and the sickly wheat on the ground of the poor field adjoining, was as straight and clean as if drawn by a surveyor. The alfalfa had undoubtedly added much nitrogen to the soil, but there were other questions involved. How much and what had been contributed by the ragweed, the decaying, natural mulch left on the field and the manure and urine of the herd of beef cattle? Was it possible too that the roots of the alfalfa thrusting deep into the minerally rich subsoils brought up from below valuable elements and trace minerals which, fixed in its roots, stems and leaves, refertilized and revitalized the worn-out soil above when the field was cultivated again and the roots and stems left to decay? There remains also the question of the bacteria closely associated with all legumes and especially with the deep-thrusting roots of alfalfa and sweet clover. Beyond fixing nitrogen could they not have had some other revitalizing effect on the worn-out soil? In any case, the fertility of the poorest field on the farm had been increased almost beyond belief within the short span of two years.

The experiment and subsequent variations of it have proven that in our part of the world alfalfa is both a poor land crop and an impressive soil-restoring crop. For us the discovery has been of great economic importance for it permitted us to utilize our steep poor land to grow one of the most valuable of crops for a livestock farm and at the same time to refresh and revitalize our land with a minimum of expenditure of cash for fertilizers. On the roadside and fence rows of Malabar there is ample evidence that the natural way of growing alfalfa in our country was the right way. Wherever seeds of alfalfa have dropped accidentally, alfalfa has sprung up and has grown as lustily as its cousin the sweet clover, without fuss or interference. Some of the plants are already five years old. The earth in which they grow was never plowed but each year it is mulched by the death and decay of the natural vegetation around it.

In the meanwhile, Dr. Borst of Zanesville announced, after seven years of tests, the results of his own experiments, conducted far more conclusively and scientifically than our own fumbling efforts. Dr. Borst, as a scientist, knew what he was driving at; we only observed results and drew our own conclusions from them. Dr. Borst proved beyond a doubt that, in our climate and in soil reasonably related in character to our own, alfalfa was a poor land crop and as much at home on poor soil that had been given lime and phosphorus as its cousin, the sweet clover. He has succeeded beyond any doubt in raising fine crops of alfalfa on miserably poor, hilly land merely by disking weeds, wire grass, poverty grass, broom sedge and other rubbish and sowing alfalfa on the trash mulch after application of lime and phosphate. His discovery is of immense value not only to the farmers of Ohio but of much larger sections of the country where for twenty-five years or more farmers have been persistently instructed in exactly the wrong way to grow alfalfa.

The benefits of a mulch on any field of clover or alfalfa or mixed hay have been proved on our previously established hay seedings where the straw, falling behind the grain combine, was left on the fields. This action was taken originally to leave the straw on the fields to decay and to be plowed under eventually in order to increase the humus content of our worn-out fields. That result was achieved but we soon found that the straw mulch benefitted the hay crop as well by keeping the soil moist and cool during the hot months and by acting as an insulating layer to keep the ground frozen all through the winter and thus prevent the destructive heaving process. Not only were our seedings more successful on fields where the straw was left, but the hay yield was greater and the length of life of the existing crop as productive hay was greatly increased.

Leaving straw on the fields meant of course that we should be short of bedding for the livestock, but that shortage we managed to overcome by doing custom work among our neighbors with our pickup baler. By baling straw on shares we brought straw onto the fields rather than selling it off them as had been done in the past. Nor did we feel that in the process we were robbing our neighbors for in almost every case straw piles after threshing were either burned or allowed to rot, although good farming practice could have put it back on the fields or worked it out through the barns in large quantities in the form of manure. This, however, was not the traditional practice and all but a few of our neighbors regarded the redistribution of straw over the fields after harvest as a waste of time and a foolish practice. That tradition accounts for the waning humus content

and fertility of many farms in our country and elsewhere in America. By baling the straw we made it more convenient to handle and actually induced neighboring farmers to use more bedding and work more straw by way of stables and manure spreaders onto the fields.

The value of mulch in gardens has long been recognized by good gardeners both of flowers and of vegetables. It is valuable not only as a means of producing better and healthier blooms or vegetables, but is actually a great saver of labor since it does away largely with cultivation by hoe or cultivator and if used thickly enough actually smothers out weeds. Growing tomatoes under mulch is an easy and profitable practice known to most amateur gardeners. Straw mulch on strawberries serves not only to keep the fruit clean; it has perhaps a more important purpose, until lately overlooked or regarded merely as incidental—that of keeping the soil about the roots of the berries cool and moist and loose, thus producing more berries and berries of better quality. The natural habitat of the wild strawberry is on partly shaded banks heavily mulched by natural accumulation of leaves, and its cultivated and highly developed cousin, the commercial strawberry, has not yet come to like or to tolerate hot, dry, bare earth about its roots.

Since we have available large quantities of manure in many forms and very often hay or straw left over at the end of the feeding season, we gradually extended mulch culture in the communal garden at Malabar to crops such as lettuce, broccoli, celery, peas, carrots, cantaloupe, sweet potatoes and other common vegetables. The results were the same in every case—that productivity, quality and flavor were all improved.

The answer lay beneath our own eyes and feet. On lifting the mulch on a day during the hot, dry weeks of August, the soil beneath was found to be cool, moist and loose from the surface all the way down into the subsoil—far looser and more open to the thrust of roots than any soil worked by hoe or cultivator. More than that, the moisture made available to the plants chemicals and elements which are not available to them in hot, dry soils, and encouraged the natural processes of decay and the growth and increase of the bacteria which promotes that process. Also beneath the mulch there was always a notable population of earthworms where in dry, hot, cultivated soil there were none. It is probable that the earthworms, moving upward and downward between the subsoil and the topsoil in which the vegetables thrust their roots brought with them, in gizzard and body, minerals and trace elements from the almost inexhaustible supply in the glacial subsoil into the topsoil where they are constantly being consumed or leached out, especially by the action of heavy rainfall on bare, exposed, *cultivated* soils. Naturally, on the mulched portions of the garden, there is no runoff water and no erosion whatever. As a supplementary benefit, all the organic material used as mulch is left on the soil to be plowed under or chopped into the soil in the following season, thus increasing greatly the humus as well as the nitrogen and mineral content. The process of mulching is not only beneficial in an immediate sense but for its great cumulative value season after season.

The virtues of mulch were evident to us not only in the natural process we used in growing alfalfa and in the vegetable gardens but in many other instances of farming and horticulture. In the very first year of the operations at Malabar, a plantation of red raspberries was established as part of the self-sufficiency program. They were put out in rows and clean-cultivated according to the traditional method in a part of the vegetable garden. The soil was free of weeds, of mulch, of surface humus, and from the first the raspberries were sickly and unproductive and subject to attack from insects.

We were on the point of giving up the cultivation of red raspberries altogether as not worth the labor of cultivation, spraying and dusting, when the idea occurred to me that the whole method of intensely *cultivating* raspberries was idiotic. In our country the wild black raspberry abounds but it grows never in a bare cultivated field but only in shaded fence rows and on the borders of the woods where the canes are partly shaded and the earth covered by a thick mulch of decaying leaf mold beneath which the soil is always loose, moist, cool and rich in humus.

Taking the habits of native wild raspberries as a model we made a new plantation in the semi-shade of the old orchard behind the Big House, less than five hundred yards from the old cultivated sickly plantation. When we came to the farm the old orchard had been cultivated in row crops and was badly eroded. The soil itself had been in much worse condition than that in the old garden where the original sickly plantation was made. During the first year of operation, we managed to get a good grass cover by sowing the eroded orchard with Ladino clover and orchard grass. Those then were the conditions of soil when we made the new raspberry plantation in the old orchard.

No preparation of the soil whatever was made. Holes were dug in the orchard grass sod and the canes planted in them and the whole plantation mulched heavily with barnyard manure. Most of the canes were fresh from the nursery but a few from the sickly plantation were used to fill out the last rows. We knew the risk of mixing the sickly canes from the old plantation with the healthy, uninfected new ones from the nursery but, in a way, that was a part of the experiment. We wanted to see what would happen.

The case history has been startlingly successful. No hoe or cultivator has ever touched the new plantation in the orchard. Once a year it has been mulched heavily with barnyard manure. Beyond that and the slight task of cutting away old canes there has been no labor expended on the plantation, not even any dusting or spraying.

From the first the plantation flourished. It is now in the fourth year of bearing and it is impossible to produce more raspberries on the same amount of ground. There are no weeds, for the mulch and the rank growth of the berry canes has choked them out. The new uninfected canes from the nursery have never become diseased from the canes brought in from the sickly plantation. What is perhaps even more remarkable, the new shoots springing from the roots of the sickly canes have thrown off all disease and are as healthy as their neighbors. The two plantations, the sickly and the healthy one, are visited by the same birds and the same bees from the thirty hives near by.

The old sickly plantation we have kept on merely as a check patch and contrast to the mulched but uncultivated and healthy plantation in the half-shady orchard. Because the old plantation had no value save as an exhibit of how not to grow raspberries, we ceased to cultivate it and in three years it grew weedy and accumulated a mulch of its own from dying and decaying vegetation. But the remarkable thing about the old plantation is that with its abandonment to the *natural* process of growth, its health has improved and each year in its half-wild uncultivated state it displays a little more vigor and produces a few more berries and new canes.

I am not, of course, suggesting that it would be profitable to grow weeds along with raspberries. I am only suggesting out of a rather startling experience that man, by following the *natural* process and perhaps accentuating it (by heavy mulch and the fertility values of barnyard manure) can save himself labor and produce bumper crops of raspberries. For the second plantation we did nothing but reproduce the natural conditions under which the native wild black raspberry flourished, with the added stimulus of the nitrogen, potassium and phosphorus contained in the manure. Beyond that, of course, we came into the realm of more or less undiscovered values of the bacteria, the hormones and the vitamins contributed by the barnyard animals themselves out of their own physiological processes. There is also to be considered the cool temperature and the loose, moist condition of the soil surrounding the roots of the raspberries in the healthy plantation.

And there is the question of earthworms which exist in large quantities in the same loose, cool soil. They are in some part attracted and their growth stimulated by a considerable amount of lost or undigested grain in the manure which in the natural process of fermentation turns to sugar and provides for the worms a rich diet.

A neighbor of ours, Cosmos Bluebaugh, one of the best farmers in America and famous throughout our state, who has large commercial plantations of black raspberries, now grows them in the same fashion with infinitely less work and much greater health and production. He discovered the virtues of the mulch system because he is a "live" farmer and noticed that where shoots from his cultivated berries found their way into a fence row and "went wild," the canes were healthier and more productive and that with them spraying or dusting for insects or disease was unnecessary. After a few experiments he went over to the "natural" method of growing raspberries once and for all. He no longer cultivates his berries. Twice a season the weeds and grass between the rows are clipped and left on the ground as a mulch. This has kept the earth cool and moist and has checked absolutely all loss of soil or water by runoff. But, as in the case of alfalfa, both our neighbor and ourselves had been taught to grow raspberries in exactly the wrong way and at a cost of much labor, of insecticides and pocketbook. Nature's way was much more simple and effective.

And then there is the case of the two plum trees which were planted about fifty feet from the house and not more than twenty feet apart where in season one could pick off a greengage to eat in passing by. Both trees came from the same nursery; both were greengages and both were planted in the same soil at the top edge of a steep well-drained bank. In the second year one of the trees (number one) showed a remarkable growth of two or three inches more at the tip of all its branches than the other tree (number two). The same record of number one continued in the third year. It far surpassed the other (number two) in growth and vigor. In the third year I stumbled upon what appeared to be the answer.

I was putting in jonquil bulbs in the heavy grass sod of the bank one evening in September. It was hard work for the sod was heavy and the ground dry and hard. Each thrust put a strain on the wrist. Unconsciously and without observing my progress as I planted I worked my way toward plum tree number one. Suddenly when I was about a yard from the tree the trowel entered the earth with the greatest ease, burying itself up to the handle from the force I had been using elsewhere on the hard hillside to make an impression. It was only then that I discovered how close I was to the flourishing tree. The discovery I made next was a simple one.

At some time when we were making improvements in the garden, someone had heaped clumps of unwanted sod lazily about the roots of the newly planted tree (number one). As the sod at the bottom of the heap decayed, fresh grass grew on the top so that the accumulation about the roots of the tree passed unnoticed, save for the evidence of its presence in the accentuated and vigorous growth of the tree above. It was only now, two years later, on that hot, dry August day that I discovered the accidentally accumulated layer of mulch. Beneath it the soil was loose and very moist. The finest of roots could push their way through and the moisture made available whatever elements of growth were present. An investigation of the earth beneath the slow-growing tree (number two) where there was no mulch, revealed earth as hard and as dry as the rest of the sun-beaten bank.

In this case the fertility elements of barnyard manure were wholly lacking. The mulch was pure sod yet its effect was strikingly noticeable. Here in the small area scarcely a dozen feet away beneath tree number one the earth revealed earthworms. A foot away in the dry, hard unmulched soil there were none.

Of course the value of mulch to orchardists has long been recognized. Our friend Bluebaugh who has one of the finest orchards of peaches and apples I know anywhere, does not use either orchard cultivator or ordinary mulch of straw and old hay; he uses his best alfalfa and feels that he makes money by doing so, not only from the ordinary beneficial effects of mulch in looseness and moisture of the soil, but from the contribution of the added nitrogen from the highly nitrogenous alfalfa. Only last summer, we used some of a weather-spoiled hay cutting from a field of mixed alfalfa and Ladino as a mulch on our small commercial orchard. Bob took time out during the busy haymaking season to distribute a couple of bales about the roots of one tree. He did this only in July but the effect even two months later was striking. The single tree mulched with alfalfa and Ladino stood out dark green and vigorous from among the others on the hillside. The difference was noticeable at a distance of five hundred yards or more.

We are able today to carry about three times the number of cattle we were able to carry on the same fields when we first came to Malabar. A large part of this is due to the application of lime and phosphorus, but again one of the important elements is mulch and humus. Both bluegrass and white clover, which are the basic plants of our pastures, like most plants flourish best in cool, moist soil. This is especially true of bluegrass which has acquired the bad reputation of going into a dormant state, sulking and refusing to grow at all during the hot weeks of mid and late summer. It is a reputation which I believe is not altogether deserved. It has been imposed upon the bluegrass largely because of careless and unintelligent treatment.

In the past and even today, the average farmer chooses for his permanent pasture the land which he finds the least valuable on his farm. Mistakenly, he does not regard pasture as a crop, but usually as wasteland where he can turn his cattle in summer and let them feed themselves. In some parts of the country there are permanent pastures more than one hundred or one hundred and fifty years old which have been neglected and overgrazed every year of their existence with the result that they become dormant and unproductive during the hot months and gradually are overrun by ironweed, thistles, sumac and wild thorn.

Gradually, through years of draining calcium and phosphorus from the soil—(for animals do not return to the soil as manure and urine *all* the minerals they consume but carry off quantities as bone and flesh)—both the quantity and quality of the pasture declines. But another element equally important in the decline of productive pasture is often overlooked and that is the gradual destructive loss of humus and mulch. Through overgrazing, no grass is left to wither and decay and add its contribution of humus and mulch to the cycle which produced our deep prairie soils and our best virgin pastures. With the natural cycle destroyed, the roots of bluegrass and other grasses gradually become bare or at best remain covered thinly by bare, hard, dry soil. Under such conditions, the bluegrass goes into a dormant stage until the cool weather and rains of the autumn season start it to growing once more. This means a loss of good and valuable pasture to the farmer for a period of from six to eight weeks during the summer months, a loss represented in dollars and cents, in lessened milk production for the dairy and the beef breeding cows with calves, and a loss rather than a gain in the weight of beef stock.

The bluegrass and other grasses cease to grow not because it is their habit to do so during the hot months, but largely because an artificial and uncongenial condition is produced by bad farming practices.

One of the worst of traditional farming practices, now happily dying out except in the most benighted and poverty-stricken areas of the South, was the burning over of pastures in the winter or spring of each year. This was done in the shortsighted and ignorant belief that the burning made clean and earlier pasture. The illusion was created by the almost unhealthy greenness of the first grass on burned-over land. It was unnaturally green because the potassium through the ashes of the burnt-over dead grass was released all at once instead of gradually through the natural process of decay. After only a year or two of the burning process all humus and mulch was destroyed and the tender roots left bare to wind and snow and sun. The grass went dormant at the season when it was most needed and eventually died and was replaced by poverty grass or the coarse and unwanted Johnson grass. In the West and Southwest, millions of acres of the richest natural legume and grass-grazing areas in the world have been destroyed by overgrazing and burning over and the destruction of all natural humus and mulch.

Since our livestock is in the fields for more than six months of the year, we have always regarded our pasture as one of the most important crops in the farm economy. In relation to the production of milk, beef, wool and mutton, our pasture land today represents, as a crop, a greater value per acre than any corresponding acre of corn, wheat, soybeans or other cultivated crops, save perhaps alfalfa alone.

It was not so when we came to the farm. Save in the bottom lands, the pasture was thin and in places there was no bluegrass or white clover at all, but only wire grass and poverty grass. All the bluegrass went dormant for a period of from six to eight weeks. Even in the bottom pasture, where the amount of underground moisture was much higher, the bluegrass stopped growing during the hot weeks.

Almost at once, we started treatment of the worn-out pastures by the application of ground limestone and phosphates. Where the pasture was poorest and we wanted quick action, the more expensive hydrated lime was used. The bald spots covered with poverty and wire grass were given light coatings of precious animal manure as well and Dutch clover and Ladino clover were seeded. Slowly, as the calcium and the phosphorus began to be available, the clover began to grow and as it restored the nitrogen to the soil, the bluegrass moved in. The manure not only added fertility, but the decaying, leached-out straw provided mulch for the young seedlings. After two or three years, the effects of the treatment became evident everywhere in increased and thickened growth of both bluegrass and white clover.

But another practice contributed enormously to the restoration of our pastures; that was the practice of mowing the pastures twice a summer, once in late July and once in early September. In theory, the practice killed weeds and brambles and promoted a new growth of young and succulent grass. Actually, the greatest benefit of the mowing process was the gradual building up of a layer of mulch and humus through the accumulation of the decaying clippings. Each year, as the layer increases, we have more pasture all through the weeks when once the bluegrass ceased growing altogether. It was not the clipping process in itself which produced the new growth, it was the steadily increasing layer of mulch-humus which insulated the soil from wind and hot sun and kept the ground cool and moist and loose. As with the alfalfa and clovers, it insulated the earth from varying changes of temperature in winter and prevented "freezing out." We had done no more than to restore the natural cycle in the life of the bluegrass.

Today we are able to carry three times as many cattle and sheep on the same ground as when we first began the treatments and during the hot, dry months there is always a supply of fresh and succulent grass, not so much as in the growing seasons of spring and autumn, but a considerable growth where we had none before. Our cattle grow and put on weight steadily throughout the summer and the milk production has been kept up even during the fly season. The beef cattle come off pasture looking as if they had been fed on grain and the finishing-off process is shortened by many weeks of feeding and labor and countless bushels of corn.

Treating pasture as a valuable crop has paid big dividends in all directions. In the future, we shall have even better returns, for not all the pasture has yet reached that peak which we hope to maintain simply by restoring the natural processes in the growth of grass. The struggle to find adequate labor in wartime has held back the whole process of restoration.

The greatest lack we found in the run-down farms which went into the making of Malabar was the lack of humus or organic material, and the same condition holds true for most of the worn-out agricultural land in America. Even in an agricultural area so fabulously rich as the delta region of Mississippi, crop yields have been declining through the steady decrease of humus. I had a letter not long ago from one of the biggest planters of the area saying that he observed that his soil was slowly becoming hard and impervious. The reason, he was certain, lay in the fact that for a hundred years or more, the only organic material returned to the soil had been the hard woody stalks of the eternal cotton. Our problem was to return to the soil as rapidly as possible and in the largest quantity possible this missing humus. Without it, the most expensive and concentrated commercial fertilizer would have been of little value. We were eager to bring up the productivity of the land, both pasture and cultivated land as rapidly as possible, but we were also determined to accomplish this by a system which recognized the fact that the average farmer was forced to grow a crop from his land *while* he was restoring it since there were always interest, taxes, seed and fertilizer costs to be paid. So the problem resolved itself into how to restore land while producing one crop a year off it which would as far as possible pay the expenses of the upkeep and the process itself.

We found a formula by which we are able to build approximately one inch or more of good topsoil a year *while* producing crops from the land. To this system Mr. Faulkner's theories made an important contribution. It is a system in which *both* trash-farming and the use of the moldboard plow play their parts. There is in it nothing unusual and nothing which is not known to any good "*live*" farmer. We have simply put the elements together in a special pattern by which the maximum results are obtainable. It is a concentrated, high-pressure system, which is in no sense a shortcut that would in the long run be profitless. It employs the methods by which Nature originally built our topsoil, simply speeding up the process about a hundred thousand times. I set it forth here because we have tried it and succeeded in building six to seven inches of good topsoil in five years while growing crops on the same land.

For the sake of simplicity I shall take one field, or rather one contoured strip 150 feet wide and about an eighth of a mile in length.

This strip was a part of the farm. The whole of the field represented a problem for it consisted of low hills or mounds with the slopes running in all directions. Nowhere, save in the low hollows where the topsoil from the higher parts of the field have been deposited, was the topsoil more than two or three inches deep. In spots, there was none at all. Below this lay the glacial moraine, a mixture of clay, sand and gravel which in itself was good potential subsoil but which was completely devoid of humus or organic material of any kind. It was what we at Malabar call "dead soil," without earthworms, bacteria or life of any sort of the kind which flourishes in the presence of humus.

When we took over the farm, the whole of the rolling field was treated as a *square* field so that there were always parts of it plowed up and down hill. When planted in row crops, the furrows between the rows each became small gullies, carrying off water and topsoil from the crests and slopes into the hollows, or off the fields carrying flooding rains into the creek near by. The fall before we took over, the field had been seeded to wheat by a predecessor and the wheat drilled up and down hill. Even with wheat as a cover crop, the roots did not hold the melting snow and the runoff water from rains, since the check rows ran up and down hill. Wherever the slope was steep, the surface of the earth was covered with small gullies, none more than two or three inches deep, but each one carrying off water and topsoil. Save in the hollows and low spots, the wheat was thin and poor.

In the spring, we seeded the whole of the field with a mixture of legumes—mammoth and alsike and Ladino clover. During the winter we limed as much of it as we were able. Part of the wheat was not worth combining but where there was wheat worth harvesting we combined it and left the straw on the ground. The legume seeding was spotty. Where we had limed, the mammoth clover made a good growth and in the damper hollows, where the good topsoil had accumulated, the alsike did well; but the Ladino seemed to adapt itself to all conditions. It grew in the damp, low, rich ground, on the bare crests and slopes, on the ground which was limed or not limed. Over the whole of the field, there was a fair cover of legumes, but not worth too much as hay.

In the autumn of that year, Hecker, our friend from the Soil Conservation Service, helped us to strip and contour the rolling field as effectively as possible. It was clear that our first job was to stop runoff water and erosion. Thus strip number one came into being. It ran the whole length of the long field most of it across a fairly steep slope. At the time it was laid out, the worn-out ground was covered by a thin seeding of mammoth and sweet clover, alsike and Ladino, with part of it covered with the thin straw left behind the combine, and part with the whole wheat which was too poor to combine still standing. There were a good many weeds of all kinds.

That autumn, or rather early in September, we plowed the whole of the strip. The season was dry and the ground, devoid of any humus, was hard, and the plowing was again, fortunately, a "poor job" for the slope and the hardness of the soil prevented the plow from turning over and *burying* the rubbish and the legume seedlings. After it was disked, the result was a rough field with rubbish, legume seedlings and wheat straw chopped up together. Into this we drilled wheat simply to get strip number one in shape for further restoration work. Although the season was dry, the wheat came along beautifully and many, perhaps as many as 60 per cent of the legume seedlings of the new seeding survived and grew. During the winter the whole of the strip was limed and given a thin coat of barnyard manure.

Several things were noticeable during the winter. (1) With the wheat drilled on contour instead of up and down hill, there was no tendency on the slope for the water to create small gullies, and none appeared. (2) The field, having been *roughly* prepared, was filled with hollows and ridges and covered with lumps of earth the size of a hen's egg or larger which disintegrated to powder-fine earth under the freezing and thawing action of the winter. All this served to collect and hold the water and prevent runoff or erosion. (3) Because the plowing job had been a "poor one," the soil was not bare. Chopped up into it were the weeds, old straw and legumes, some still alive and growing. This condition too served to hold the rain water and check erosion. But the most notable effect was the feel of the soil itself. When you walked over it, it was *springy* under the feet from the presence in the soil of decaying straw and rubbish which had been mixed *into* the soil instead of being buried. When one walked over the field the year before, it was like walking over a cement pavement from which all the water ran off. During this second winter the surface of the whole field was like blotting paper. It held *all* the rain and checked *all* erosion.

The effects showed on the wheat yield. From the very beginning of spring it was evident that despite the fact that we used no more fertilizer than the farmer had used the preceding year, we would have a pretty fair crop of wheat on land which the year before had produced a crop not worth combining. The season was an average one and the wheat averaged over twenty bushels to the acre, a poor yield but approximately 300 per cent more than the preceding years. Certainly we owed a large part of the increase to keeping the water where it fell instead of permitting it to run off. But the looseness of the soil with the chopped-up organic material in it also permitted better root growth and the added moisture made the fertilizer used available to the plants and about twice as effective. The barnyard manure, although much of it had been left in the barnyard by our predecessor and was badly leached, contributed its good effects in terms of all the valuable known and unknown elements which animal manure contains. And there was the nitrogen being given off slowly by all the decaying weeds, straw, etc., mixed into the soil. The legumes which had survived the "poor plowing" and disking continued to grow with the wheat, supplying nitrogen to the soil. These, particularly the sweet clover, grew high and caused some trouble in combining, but not enough to outweigh their great value.

The straw was left in the field together with the legumes, principally the tall sweet clover, which had been clipped in the process of combining. In the autumn of that year (again early September), strip number one was put into cultivation. This time it was not plowed but ripped up with a tiller and disked once. The tiller was equipped with alfalfa "knives" which tear through the soil but do little damage to alfalfa or sweet clover roots. Despite the disking which followed, most of the original legume seeding still survived and grew. This time, there was chopped into the earth but not plowed under and buried, a second year of straw, the remains of the clipped legumes left behind the combine and what was left of the application of barnyard manure. The surface of the field still remained rough, which was what we desired. This time, all the material was chopped into a depth of three to four inches and for the first time there appeared the elements of topsoil.

Into this was drilled rye and vetch seed which, in the stored moisture of the blotting-paper surface germinated almost at once and made a quick growth. By the time winter came there was a thick green mat of rye and vetch over the whole of strip number one.

In the spring, the strip was ripped up again with a tiller and disked again and this time sown to oats. We did not sow the oats early as is the proper rule if one is to catch the cool weather and early rains of spring. There were two reasons for the delay—one, that there was more important work to be done and as we did not expect a great yield of oats from strip number one, we let it go until last; and second, we wanted the rye and vetch to make as much growth as possible before disking it in, in order to get as much organic material as possible.

We rough-disked the field and sewed the oats nearly a month late, when the rye and vetch was about two feet high. Because it was thick, it took a good deal of fitting, but when we were finished, the soil for a depth of five to six inches was a mixture of earth and chopped and decaying rye and vetch stalks and roots. Into this, the oats were drilled, with no nitrogen fertilizer and the usual amount of potassium and phosphorus. We were right in holding back on the nitrogen; the legumes, the decaying rye and the vetch supplied all the nitrogen and more than was needed. The oats grew rankly, too rankly in spots where the original legume seeding had been thick. They showed no bad effects from the late planting because the heavy mulch of decaying rubbish kept the ground both cool and moist. Before the oats appeared above ground, we seeded the strip to sweet clover and, owing to moisture and mulch, we secured a fine even stand of this tricky but valuable legume which is sometimes difficult to get started.

The oats yield was startling, from four to five bushels per acre above the average yield in our county which is not the best oats country. The straw again was left on the strip and the sweet clover seedlings came through it easily and grew far ranker than any sweet clover seedings we had ever made; no doubt because the whole of the soil, filled with decaying vegetation, beneath the mulch of oats straw, remained loose and moist throughout the summer.

That winter, the stems of the sweet clover seedlings froze back and fell to the earth to add their nitrogen to the already thick layer of mulch, and the following spring we were ready for the third stage. By now, without plowing, we had from three to four inches of pretty good topsoil and at least another inch or two of oat-straw and sweet clover mulch lying on the surface. In late May of the third year, when the sweet clover had reached a height of about two feet, we put the plow to work and turned the whole layer of new topsoil, decaying mulch and the heavy growth of sweet clover completely over or upside down, plowing deep to bring up from two to four inches of the buried soil and topsoil. This was disked heavily to mix it as much as possible with the newly created topsoil we had buried. The strip was planted to corn with soybeans sown in the rows with the corn, the whole to be cut for silage. We did not, of course, measure the crop yield in ears. We planted hybrid DeKalb 604 and the results were remarkable, thick deep green corn with two big ears on two out of three stalks, with succulent soybeans growing three to four feet high. The strip filled our biggest silo and part of another.

The best comment came from our neighbor, Charley Schrack, who said, "I've lived next to that field for fifty-four years, since I was born, and I never saw on it as fine a crop of corn."

Once we had turned the soil upside down, the same process of small grain rotation with intensive planting of legumes was followed. Next spring we shall turn the top layer upside down in turn, plowing deep to bring up subsoil if possible. At the end of five years we had built six inches of good topsoil, as good—indeed I am not certain that it is not better topsoil than Nature had spent several thousand years in creating. For the mineral balance is better than that of most topsoil produced by Nature especially in areas once covered by forest. We shall husband it carefully in the future and see to it that we do not take off more than we put back.

The system we used was an intensive but not unreasonable one, and after the first year succeeded in raising a profitable crop each year. Both nitrogen and organic material were restored to the soil in large quantities, as well as the minerals brought up from the subsoil through the roots, stalks and leaves of the deep-rooted legumes. To all this, lime, potassium and phosphorus and trace minerals were added artificially to restore them to the leached-out soil.

I do not believe that the rebuilding of topsoil could have been accomplished so speedily if we had plowed instead of using the trash-farming methods advocated by Mr. Faulkner. There are many reasons for this belief, principally that by plowing and burying manures, stubble and straw, much of their value, both in nitrogen and humus, is destroyed. The surface rubbish of all kinds, when *plowed* under by a good job of plowing is compacted into a narrow tight layer pressed down by the weight of the earth as well as that of the implements passing over the ground during the process of fitting. During the heat of summer, the compacted mass does not decay, but ferments, creating acids which actually eat away into nothingness the organic material which should be converted by the slower process of decay into humus. The process also locks up nitrogen instead of releasing it and by the acid fermentation and the absence of air, destroys the bacteria so valuable to plant growth or checks their action in the *natural* process of decay.

When the same amount of manure and trash is chopped into the soil, there is no heating or fermentation. The organic material is converted into a maximum of humus, slowly, through the natural process of decay. The soil remains loose and open, springy as you walk over it, open to the action of bacteria, of air and of the water which it entraps. In other words, plowing is the unnatural process and trash-farming the natural one.

Long before the experiment was made, I had seen in the state of Indore in India the results of experiments made by Sir Albert Howard during his long stay there. The basis of Sir Albert's process of rebuilding and revitalizing soil was that of composting, not only animal and green manures, but all available rubbish and refuse. It was a process which began actually in the stables themselves, where an effort was made to preserve *all* animal secretions and to keep alive and promote the increase of bacteria. Toward this end, Sir Albert had six inches of absorbent clay laid down in the stables to absorb all liquids. This clay was removed periodically and placed in the compost heaps along with the manure, straw and disintegrating rubbish. During the process of decay, the compost heaps were turned by coolie hand labor from time to time, to prevent heating and fermentation. When the composting process was completed, the residue was removed to the fields where it was mixed with the worn-out Indian soil and produced prodigious results in health, vigor and the productiveness of plant life.

The Indore process of Sir Albert Howard produces perhaps the ideal fertilizer, containing not only chemical elements necessary to plant growth, but also the residue of *all* animal secretions and billions of active and stimulating bacteria. Beyond that, it is in itself *living* humus and as such absorbs and preserves moisture and helps to achieve the *maximum* result from any chemical fertilizer used in the field.

Such a process and a result would have been invaluable to us at Malabar, as it would be to any farmer, but the expense of time and labor made it impractical and indeed impossible. The coolies employed in Sir

Albert's process received approximately twenty cents a day—a wage that was nonexistent in America or even in Europe. The lowest wage in our Middle Western country for unskilled farm labor is approximately ten times as great. Throughout much of the period during which our own experiments were carried on, farm labor at any price was very nearly unavailable. We had, therefore, to find some adaptation or variation of the Indore process.

So far as the stables and feeding barns were concerned, we simply substituted ordinary sawdust for the absorbent clay employed by Sir Albert. This we used as bedding twice a week in layers between the straw. The sawdust absorbed all urine and glandular secretions and the urine began almost at once a process of breaking down the tough cellulose structure of the sawdust, thus hastening the natural process of decay into humus and available plant food. The stables were cleaned periodically and the manure spread over the fields, without the loss of any essential elements. During the period it remained in the stables, no leaching by weather occurred at all and the liquid content prevented heating or fermentation.

Once in the field, the manure was not plowed under to ferment and form destructive acids. It was stirred into the soil along with whatever rubbish, weeds or green manure lay on the surface. During the processes of cultivation, or even without cultivation, the mixed soil, rubbish and manure was left loose and open, available to air, water and even light so that the essential processes of decay which took place in Sir Albert's compost heap also took place in our impoverished topsoil. Actually, we did no more than transfer the composting process from the compost heap into the topsoil of our fields. We were growing crops in a living compost heap, with *no* additional costs either in time or labor.

I have not made tests as to the different amounts of nitrogen released respectively under the two methods. I have not counted the abundance or lack of bacteria at work. I only know by the evidence of the plants growing in the trash-farmed soil. They *like* it and are healthy and productive. I have seen the topsoil grow under my eyes and beneath the tread of my feet. I have had good farmers and scientists comment on the springiness of the soil which five years earlier was little more than cement devoid of humus. I have heard neighbors comment on the darkened *color* of the soil. I have seen the earthworms increase from none at all, save in the damp topsoil-filled hollows of the field, to a very busy population distributed evenly over the whole of strip number one. I have seen the population of bumble bees increase from virtually none at all to numbers so great that their humming could be heard at a distance of fifty yards from the field. I have seen pheasants and rabbits and quail return to land they had deserted because it was dead. In short, I have seen the stripped, worn-out soil come back to life and productivity.

The system used on strip number one has been used with some variations on other fields and always with similar results. The increase of humus proved another fact of great importance—that once the content of organic material is sufficiently increased, erosion, even on steep slopes, is checked, for the soil acts like blotting paper and absorbs all the water which falls on it. In other words, once we restore the natural humus content of the soil in the whole field crossed by strip number one, it will no longer be necessary either to strip or contour it. So long as that humus content is kept high there will be no runoff water and no loss of topsoil.

The processes which we came to use in the restoration of organic material and the building of topsoil were no more or less than the processes used by Nature herself or in the conduct of any well-managed compost heap. The natural process was hastened by the added concentration of organic material in the form of green and animal manures and by the fact that all this bulk was not plowed under, buried and compressed but mixed loosely into the soil where the crops were growing. The compost heap, in other words, was transferred from the pit or shed to the topsoil itself. We had built upon the natural formula of life and vigor growing out of death and decay. All we did was to speed it up several thousand times. Thus we achieved a kind of resurrection by which dead soil rose from the dead and became alive once more.

OF BOXERS, A MONGOOSE, SOME GOATS AND SHEEP, A BULL AND A DUCK

(To be skipped by demon housekeepers and people who don't like animals.)

THE BOXERS BEGAN WITH REX, A KING AMONG DOGS. HE arrived one afternoon at the house in France, big, golden brown with a muscular body, a black face on a snub bulldog head, a broad chest, an appearance of great ferocity and an air of great dignity. He was the gift of a friend and he had been spending several days with her twelve Norwegian elkhounds in her house near Chantilly. Now elkhounds are not exactly lap dogs and they can fight like demons and they resented the presence of the newcomer from Germany. There were three days of fighting and in the end, Rex the boxer won, if not the victory, the right to go his dignified way, unmolested. He arrived at our house with the sense of victory still strong in him.

It was not like having a puppy arrive in the house. That was an experience we had had countless times, mostly with Aberdeens, for we had come to have several dogs instead of merely one because when anything happened to one the sense of loss was too great. So we were always having new litters of puppies or buying new ones. Later we discovered that having more than one dog didn't make much difference because each one has its own personality and is an individual and when he dies, the sense of loss is not softened by the fact that he left behind several companions.

Rex was the first grown dog who had ever come into the house, and it would be hard to imagine a dog of more awe-inspiring appearance. I get on well with dogs but I confess that on first sight I had my doubts about getting on with Rex. He came into the room with an air of complete self-assurance. He did not wag his tail or growl or sniff about. He walked in with dignity and stood there looking at us. It was clear that we were not looking him over. The process was quite the reverse. He was a personality with great dignity who expected no nonsense and no familiarity. With some misgiving I patted his head and spoke to him. He neither growled nor wagged his tail. It was as if he were permitting graciously a favor.

That night I fed him and took him to sleep in my room. He went along graciously but with the same dignity and detachment, and in the morning when I wakened he was there on the rug beside me, awake, with his head between his paws, watching me out of his big brown eyes. It was then I think I realized for the first time what lay behind the eyes—the affection, the devotion, the loyalty, the dignity and the independence. But still he gave no outward sign, either of affection or even acceptance. When I went downstairs he went with me into the garden and while I had breakfast he sat in the dining room near me, never begging, taking no notice of the fact that I was eating. He had been a show-dog, and like all boxers was a gentleman. I did not know then that he had never really had a master or lived with a family and that he was desperately hungry for a master to whom he could attach himself.

For three days he lived with us, still with the same dignity and detachment, never once growling or wagging his tail, and then on the fourth morning, when I wakened, he got up off the rug and came and put his big head on the bed beside me and looked at me with his great brown eyes. He did not wag his tail. He simply looked at me. I put out my hand and rubbed his ears and I knew that we were friends. When we went downstairs he went up and sniffed at each member of the family and did the same trick of resting his head on their knees and then suddenly he wagged his stump of a tail. He was telling us that he had accepted us after looking us over for three days. From that moment on there was never a more devoted dog. He took over the whole family. He watched the house. He was happy really only when the whole family was at home and together. Even then he kept trying to round us up and keep us all together on the terrace or in the library or the salon. We were his responsibility. When part of the family went to Paris for the day, he would sit all day listening for the sound of the car, an old Peugot, which he could distinguish from every other car passing the house. When he sprang up and barked and ran to the garden gate, you knew the family had returned safely and he would be happy again, because we were all together under his watchful eye.

As the days went by his character softened and at times he would become almost demonstrative, although he never lost his dignity and he was always a little ashamed of displaying any emotion. I think he must have had an unhappy experience as a puppy, of being shipped about at dog shows and finally sold and put on a train and shipped to a foreign country and sent to a strange house filled with strange dogs where he had to fight to assert his domination. I think he had come to distrust all people and perhaps all dogs. That is why he had looked us over until at last he decided that we were all right and that he had no reason to distrust us. I do not use the word fear for there never was in Rex in all his life any sign of fear.

In the house at Senlis, he had two rivals, one of whom he dominated, the other he never managed to subdue although she weighed less than one of his own big paws. The first was the grandfather of the Aberdeens—a grizzled, tough old Scotty called Dash. Dash was both a tramp and a Don Juan. All the town knew him and certainly every lady dog in the town was acquainted with him. He would fight dogs of any size. I have seen him lying on his back snapping and biting at a sheep dog four or five times his size until the bigger dog would yield the day.

There was a square in the town where all the stray dogs had a habit of gathering, perhaps because in the square there were three butcher shops with open fronts and the butchers had a habit of throwing scraps into the street for the dogs to fight over. When any of the family went up the hill to the market, Dash always went along. He liked any excuse for going to town. And when you turned into the square Dash would rush forward at full speed into the midst of the little army of mongrels sunning themselves or scratching peacefully on the cobblestones. At sight of him they would scatter in all directions into alleys and doorways. He never failed to perform this same trick.

He was, in addition to being a tramp and a Don Juan, a good deal of a show-off. He knew all the butchers well and his showing off paid good dividends, for the spectacle of the fleeing dogs always amused them. When the square was cleared, Dash, sometimes accompanied by a lady dog, would quietly visit each shop in turn and receive scraps which he devoured in peace without any vulgar fighting.

Every family in town with a bitch sooner or later would announce that their bitch had had puppies and that Dash was the father. They always had a certain pride in the event since Dash was not a mongrel but a *chien de race*, a pedigreed dog, and that, no matter what the breed or ancestry of the female, made the puppies in French eyes, distinguished and valuable. But of all his conquests the most remarkable was that of a huge German police dog bitch called Marquise. She was the property of Picquet, the gardener, and I doubt that a more ferocious dog ever lived. She had come to Picquet full-grown and most of her life she had spent chained to a laundry delivery wagon. She appeared to hate all mankind and all other dogs. Picquet kept her attached to a kennel by a heavy chain near his house in the vegetable garden and each time anyone opened the gate, she would lunge forward growling and showing her teeth. She grew steadily more savage until, fearing for children and guests and even for myself, I asked Picquet to get rid of her.

Picquet himself, was what might be called a "natural." He was not very bright but he had a wonderful way with flowers, vegetables and animals. I doubt that anyone else living could have gotten on with Marquise. He came from the Pas de Calais and spoke the ugly half-Flemish Pas de Calais *patois*. He could wait on table, take care of the pony and the children, poach trout out of the clear-running Nonette by luring them with Swiss cheese or stunning them with *eau de Javelle*. Together he and I grew wonderful vegetables and flowers and I learned from him many a trick about gardening and farming which I hope never to forget. I think in all that world he was the only person who loved Marquise. I promised him another dog if he would send her away.

He agreed as to the danger of having Marquise about, but asked that he might keep her a little longer. She was, he said, *enceinte* which in English merely meant that she was about to have puppies. I received this news with astonishment as I knew that not only was Marquise kept chained up but that the garden was surrounded by a high wall topped with broken glass which no dog, not even a St. Bernard, could cross. The only dog who ever came into the garden was Dash who worshiped Picquet and left him during the day only on the occasions when he went to market to break up the dog *Kaffeeklatsch* in the Place Gallieni and receive his daily handouts from the butchers. Considering the difference in size between Dash and the ferocious Marquise, a mating seemed highly unlikely. I said to Picquet, "But what dog could be the father? No dog but Dash ever comes into the vegetable garden."

Picquet bowed his head and said, "*Mais monsieur, J'ai aidé un peu*" (I gave a little help).

Marquise remained until she had weaned her puppies and then went off as the companion of a forester living in a lonely house on the road to Aumont. Everyone was happy, including Marquise, for in the forest she could run at liberty. Picquet kept two of the puppies who grew up into odd-looking dogs—half-pint police dogs with smooth black coats, Scotty heads and magnificent sweeping black tails arching over their backs.

On one occasion Dash very nearly destroyed a famous dog, the handsome white poodle called Basquette, belonging to Gertrude Stein. Basquette, a big pink and white dog, trimmed and tonsored always by the best dog hairdresser in Paris, arrived with Gertrude for lunch one Sunday and Dash, after one look at Basquette, decided, I think, that the poodle was not a dog at all but some monstrous strange, unknown animal that asked for extermination; an intolerable sissy to a tramp Don Juan like Dash. In any case, he leapt to the attack and while the pampered, marcelled Basquette howled, Dash went for him. Two tables were overturned, one guest kicked accidentally in the shins by another and two others bitten before Basquette was rescued, his beautiful marcelled white coat streaked with blood.

Only once did Dash ever attack a human, and on that occasion I felt that he was justified. His victim was a pompous bore, the town jeweler, by name, Monsieur Bigué. It has often been said that bores recognize and avoid each other but this was certainly not the case with Monsieur Bigué who had chosen to marry a bore even greater, if possible, than himself. By some ill fortune, the pair had one of the common gardens which adjoined our own and they had a dreadful habit of leaning over the fence to make long, involved, formal and deadly conversations on politics, the weather, the foreign exchange. They were childless and incredibly avaricious as only a French provincial jeweler can be. Madame Bigué lavished all her affection upon a wretched little female pocket-sized dog called FrouFrou which fortunately never left the jewelry shop or house save to go into the little walled garden at the back. No virgin heiress was ever protected as fanatically as Frou-Frou. She was, I think, the only female dog in the town which Dash had not seduced. Madame Bigué would, I think, have preferred the grave to bringing Frou-Frou with her when she and her deadly husband came to work their little garden or sit in the summerhouse which dubiously ornamented it. Dash's evil reputation had long since preceded him.

For years Dash took no notice of Monsieur Bigué although he passed the kitchen steps almost daily, always carrying an umbrella and a basket. (In addition to being bores, they were a mistrustful, suspicious pair, who never for a moment trusted anyone or anything, even the weather.) Then one morning without visible provocation, Dash sprang from the kitchen steps and tore the seat out of Monsieur Bigué's pants as he passed with his basket and umbrella. Marguerite, the fat cook, rescued Monsieur Bigué and when I heard of the incident I went to wait upon Madame Bigué at the jewelry shop. She received me with dignity as I explained my regrets and said that I wished to make reparation.

All would be forgotten, said Madame Bigué (while the virgin Frou-Frou yapped in some hidden portion of the house) if Dash were left with the veterinary until it was determined whether he had hydrophobia and if I paid the doctor's bill and for a new suit of clothes for Monsieur Bigué. It just happened, said Madame Bigué, that Monsieur, although he was bound to work in the garden, was wearing his best suit, and as it was obviously impossible to buy a new pair of pants that would match the rare pattern of Monsieur Bigué's choice. A mere new pair of pants wouldn't suffice; it had to be a whole new suit. I inquired for Monsieur Bigué but was told that he was prostrated and that his leg was swollen to elephantine proportions.

The story ended happily enough. I paid the doctor's bill and bought a new suit for Monsieur Bigué. Dash, it turned out as I expected, did not have hydrophobia. I think he merely had great wisdom. I never knew whether Dash attacked the jeweler because he had heard about Frou-Frou and the seclusion in which she was kept or whether he did it because Monsieur Bigué was an intolerable bore and Dash could not support watching him pass one more time with his basket and umbrella. I only know that if Dash hadn't attacked him, I should have bitten Monsieur Bigué one day myself.

That then, is the history and character of Dash, not a dog who after ruling the dogs in his own house and indeed in all the town, would welcome the arrival of a big boxer from Germany.

On the first morning after Rex's arrival, the two dogs encountered one another in the garden. They did not growl. They did not sniff at each other. As for Rex, he walked past Dash with perfect dignity, ignoring him. For once in his life Dash did not attack on sight. His tail went straight into the air and twice he walked the length of the garden on his toes, very stiff-legged, every hair on end. He attempted to ignore the big bulldog but couldn't quite succeed. He kept watching him out of the corner of his eye, sizing him up. For once, it seemed, he decided he had met his master. Presently, with an air of the utmost casualness and dignity, he re-entered the house and went through it to his post on the kitchen steps where he could watch what went on in the street and attack any unfortunate dog who happened to pass by. As a tramp, the kitchen steps were just the place for him.

Relations between the two dogs never improved although they continued to live in the same house. When they encountered each other, Rex ignored Dash and Dash walked past Rex with the aggressive stiff-legged dignity of a very small Scot. Rex was the only dog that Dash failed to attack in the whole nine years of his life. I think it was not the size of Rex which awed him for he had fought even bigger dogs and sent them howling-down the street; it was the awful, regal manner which Dash had never encountered among the mongrels of the town.

But Rex's other rival was of a different character. She was tiny, with gray and black speckled hair which made her coat appear less like hair or fur than like the feathers of a genuine fowl. She weighed a little over a pound and had a tiny pointed head and a delicate pink nose with brown shoe-button eyes. When she was angry she would sit up, balancing herself on her tail and chatter. All her hair stood on end so that she took on a bushy appearance and the brown shoe-button eyes turned red. She had tiny brown paws which gave her the effect, when she sat erect, of a very chic lady wearing brown gloves.

George and I had found her in a market at Madras being dragged along on a string by a Pathan six feet six with a foot-high turban and a menacing appearance. One of those things happened. Obviously she did not like being dragged along and had no intention of giving in. She chattered, her eyes turned red and all her hair stood on end. There seemed only one way of rescuing her and that was to buy her. After some double talk haggling with the big Pathan we bought her for thirteen rupees which was more than she was worth.

The moment the Pathan stopped dragging her along on a string her temper began to cool. I put her in the pocket of my coat with the string still attached and she went along perfectly happily. She had to have a name and there was but one universal name for a mongoose. She became Rikky after the mongoose in Kipling's famous story.

The odd thing was that she settled in at once on George and me. Mongooses are curiously sociable animals and like to live with people, so long as their freedom is not restricted. All over India mongoose families live in peace and contentment in and under bungalows, driving away the snakes. And so, although we removed the string almost at once when we returned to Government House where we were staying, she did not run away but stayed about following me everywhere and putting her pink nose into everything.

If a mongoose has one outstanding characteristic it is curiosity of an unbelievable violence. Rikky was no exception; she had to open every drawer and box, get inside every piece of luggage to see what was there. If a cigarette or matchbox offered difficulties she would work over it until she found the secret. One afternoon her curiosity brought disaster. I had sat down, all dressed in white flannels for a visit to the palace, to write a letter and no sooner had I begun than Rikky was up on the writing desk, watching the movement of my pen. She reached out with her little brown-gloved hands now and then to touch the pen. When she had divined approximately what I was doing, she turned her curiosity elsewhere, this time to the ink pot. I pushed her away once or twice but instantly she was back again and, eager to get on with my letter, I forgot her until suddenly she jumped from the table into my lap and then I saw what she had been doing. She had been dipping her tiny paws into the inkwell and as she ran from my lap up to my shoulder she left footprints of ink the whole length of my white flannel suit.

Sometimes during the day and always at night, she would disappear. At first I was alarmed but always she came back. And then one day I discovered where she went—inside the springs of the mattress. Later on I discovered the trick and why she did it. Being an immensely active animal with a restless, lightning-like movement all during the waking hours, she slept soundly when she slept, so soundly that on occasions when I found her hiding place, I had at first the impression that she was dead. It sometimes took as much as a full minute to rouse her. And gradually I came to understand why she hid herself away when she slept. It was the ancient instinct of protection. She would never sleep in her basket but always sought a place where she would be completely hidden away, where no enemy could surprise or attack her while she was sleeping. Once while we were traveling in a small launch down the backwaters of the Malabar coast, she hid herself away so thoroughly between the floor and the hull of the launch that we had to tear apart the boat to find, waken and recover her. It cost me money for repairs, but by that time I had become so attached to her that I would, I think, have bought the whole launch rather than lose her.

She came back with us to France, most of the way in my pocket, on a P. and O. boat. I was troubled about the difference in climate and how she would survive. The moment she reached the house and garden she looked things over and decided somehow that this was home. For the first night she slept inside a big upholstered sofa but after a day of reconnoitering, she came to the conclusion that she had found a better place. I should have had no concern for never was there a small animal more capable of taking care of herself. She had in fact, chosen the space between the roof of the stable and the flooring of the attic above. There she built herself a nest of old rags and bits of newspapers, salvaged from the trash bin. She slept amid fragments of the *Figaro*, *L'Illustration* and the *Revue des Deux Mondes*.

Before I knew how well she could take care of herself, I drilled a small hole in one of the doors so that she could come in and out at will when the nights were frosty. In my stupidity I believed that she did not know when the weather was too cold for her outside. I had a small slide of wood fixed so that once inside, I could close the opening and keep her in the house. In order to teach her what the hole was for (a piece of presumption and nonsense on my part) I put her through the hole and then went inside and pushed the slide across the opening. And instantly, as she had done so many times, she made a fool of me. She sat up on her hind legs, took one look at the slide, pushed it aside and went out. Not to be baffled, I fastened a hook to the slide to thwart her. Again

I put her through the small hole and this time fastened the slide with a hook. Immediately she attempted to push back the slide and at once discovered that the trick wouldn't work. After studying the situation for a second with her tiny head a little on one side, she went to work on the hook and in less than a minute discovered how it worked. She lifted the hook and went out again.

After that experience I gave up and left the hole in the door open and then I discovered that she knew perfectly well, far better than myself or any weather forecaster, when it was cold enough to come into the house. Indeed, she became a perfect weather forecaster. If there was to be a sharp drop in temperature, she would come into the house and sleep inside the springs of the sofa. Otherwise she stayed in her nest in the stable.

She would come to the sofa on other occasions, usually when there were people for lunch or visiting dogs. Then she would thrust her nose out from under the cushions and lure the dogs toward her. When they thrust their noses into the cushions she would give them a sharp nip that sent the dogs away howling. On more than one occasion she startled guests seated on the sofa and sometimes unaware even of her existence, by dashing in the door and into the sofa between their ankles. She had a trick of showing great interest and curiosity in people who were terrified of her. Malice, indeed, was as much a part of her character as curiosity.

One of her favorite tricks was to climb inside my shirt into the sleeve and come down my arm, thrusting her pink nose and bright eyes out of the cuff to watch what was going on. She would sit on my shoulder and nibble at the lobe of my ear with her sharp teeth, gently, without ever harming me. I think it was her one way of showing affection. If I went into the garden and called her she would either give no sign of hearing me or would put her head out of a crack of masonry in the old walls and chatter impudently without any sign of coming down. But, sooner or later, she would appear, *casually*, on the garden path as if she had just happened along on her own. And she had another trick, done out of malice with every evidence of a sense of humor. While I worked in the garden with my mind far from her, she would sneak up under cover of the vegetation and jump suddenly onto my head, clinging there with her tiny feet. She never failed to get the expected result of a yell.

She was death on rats and frogs. The big river rats from the little river which ran through the garden were too big for her to kill but she had her way of dealing with them and before long drove them away altogether. I have watched her at work. She would hide and when a great rat came lumbering out of the water, she would dart out like lightning, give him a thorough clawing and biting and disappear again before he knew what was happening. When you heard wild squeals from the riverbank, you knew she was at work. On more than one occasion she killed a whole litter of young rats in the old chapel or the garage and always she left them, laid out neatly on the garden path in a row. I do not know why she did this, but it was as if she said, "See what I have done!" She would play eternally the cobra game. I would strike at her with the striking gesture of the cobra and at the fraction of a second my hand struck at her, she would leap straight upward three or four feet in the air. Never was I quick enough to touch her.

She did not eat the rats. She liked snakes and frogs and the eggs of the white fantail pigeons and had a great fondness for hen's eggs which were given her once a day. To a stranger it was always a source of amusement to watch her cope with the problem of breaking the shell. She would stand on her hind legs and with her front paws push the egg backward between her legs, in the fashion of a football center passing the ball, until it struck a stone step or wall and the shell was broken.

When she first came to the house in France her playmates were two puppies, a Scotty and a mongrel fox terrier called Albert, and a Siamese cat called Sita. With the puppies the games were all play but with Sita, there were times when the jungle emerged in both Rikky and the cat, moments when on the ground or in a tree they went for each other with teeth and claw. It was always Rikky, with her lightning quickness who had the advantage.

It was Rex, the big boxer, who aroused her jealousy. The other dogs she did not mind or the Siamese cat, but a few days after Rex arrived, she divined, I think, that he had a special place in my heart, and from then on she gave him no peace.

At first he chased her and I feared that he might do her harm. Each time I called him off, until with his good boxer character he controlled himself and adopted a new tactic of ignoring her. This only made her more malicious and she would follow him about, trying to lure him into chasing her, sometimes even running between his legs while he walked with me in the garden. If he sat or lay on the grass she would come sneaking toward him, her little eyes red with jealousy and anger, every hair standing on end, her tail suddenly inflated to the size of a fox's brush. I have seen her creep close to him and put her pink nose against his big black muzzle while he trembled all over with restraint. If he made a sudden move she was off like lightning up the nearest tree or wall. She knew that he never had a chance against her quickness. At last, unwilling to see the well-mannered dog tormented by her malice, I said, "All right. Get her!" After that he chased her whenever she began the tactics of annoying him. He never had the remotest chance of catching her and I think she enjoyed the game of being chased, knowing perfectly well that she was safe.

It was a feud, a jealousy between the tiny mongoose and the big dog, which never died until the day Rex and I left France and were forced to leave her behind.

This happened when Mr. Chamberlain went to Munich with his umbrella and rubbers to meet Hitler. On the day the meeting was announced any sensible, well-informed person knew that there was no longer any hope of averting war and the collapse of Europe. I sent my family home to America. When, at last, I listened to the counsel of Louis Gillet and returned home, I could not take Rikky with me because no mongoose was allowed to enter the United States.

I made the discovery nearly a year before when Mrs. Lehman, who loved animals, cabled me suddenly from New York—AM SENDING MY MONGOOSE TO YOU STOP NOT PERMITTED TO ENTER THE U.S. Ten days later there arrived a box and in it a mongoose. I had received the cable with pleasure thinking that Rikky would have a mate or at least a companion, but when I opened the box I had a disappointment. What came out was not a gay, wicked, malicious slim creature like Rikky but a rather heavy animal with black rings around its gray body, looking more like a rat than like Rikky. From the first it appeared to be a stupid and ill-tempered beast which, despite its apparent tameness, would bite at the least provocation. It was a mongoose all right, but an African one.

Rikky would have none of it. After circling about it suspiciously for a time, she attacked it, using the same hit-and-run methods she employed on big rats. The African mongoose had none of Rikky's quickness and after a few moments' bewilderment fled in the direction of the river. I rescued it from the water and took it into the house, thinking that at a second or third encounter Rikky might become used to it and accept its company. But it was no good. Rikky considered her African cousin a stupid monstrosity and after she had twice again driven it into the river, I gave it away to Charlotte Erikson who lived in the same town. She kept it and managed to cure some of its ill-temper, although it always remained a treacherous beast which never learned Rikky's trick of sitting on your shoulder to nibble your ear quietly and affectionately.

When life in Senlis came to an end forever and we left the old house by the Nonette, George took Rex with him on the "Normandie" straight to America since the quarantine on dogs in the British Isles made it impossible to take him with me to London.

Rikky was left behind with Picquet, the friend of Dash and Marquise who was next, I think, in her affections to me. The thought of leaving her with Picquet softened the sorrow of parting a little for I knew that, with his quality as a "natural," he understood animals. For more than a year she took over Picquet, not as a master since she never accepted a master, but as a friend. Dash was dead, having passed away of premature senility, owing, I think, to a disreputable but highly enjoyable life. The one remaining Scotty, not a very bright or affectionate dog, I left with a friend. Sita too was gone, after having become paralyzed from having too large a litter of kittens by an alley tomcat. The old life had come to an end.

But at last the mobilization caught up with Picquet and he went away leaving funds with a neighbor to see that Rikky got her daily egg. The Germans came to the house and for a time even the Senegalese were quartered there. I had news of this and worried over what had become of Rikky, but presently I had a letter passed through the Underground from the neighbor and another from a Belgian friend who was a master pilot on the "Scheldt" and was passing through Senlis as a refugee being repatriated. Both saw Rikky who thrust her head out of the crevice in the wall and then came down to see them. She seemed happy and in good health and quite able to take care of herself. She was still there when the neighbors returned to Senlis and, although no one occupied the partly wrecked house throughout the war, she stayed on there feeding on young rats and frogs or scraps and an occasional egg from the neighbors. I had a couple more letters through the Underground, principally about Rikky to tell me that she was alive and flourishing and still living in her nest between the floor and the ceiling of the stable.

She must be getting old by now for a mongoose and someday, I suspect, she will crawl into the nest of *Figaros* and *Revues des Deux Mondes* and go into one of those trancelike mongoose sleeps from which at last she will not waken. Sometimes I have wondered what the German troops or the black Senegalese thought when suddenly the strange little animal, the like of which they had never seen before, appeared in the garden. Did she climb up on their shoulders or jump suddenly on their heads. I think not, for she was both a cautious and an intelligent little beast. If she did they were probably so startled that they made no attempt to harm her. In any case she was always too quick for a pistol shot ever to end her life. I like to think of her tantalizing the Germans as she tantalized Rex. And it is a great tribute to her personality that people bothered to write letters and smuggle them out of France through the Underground just to tell me that she was alive and flourishing. But the French are like that where pets are concerned. It is one of the reasons I have so great a love and respect for them.

In London I heard only assurances, mostly from Tory friends, that there would be no war, that Mr. Chamberlain and the Conservative party had arranged all of that. "Peace in our time!" was a phrase which became more and more insipid and unreal and maddening to someone who had been living on the continent of Europe for the greater part of eighteen years and who had heard at dinner parties in London and Paris, Madrid and Rome the talk of the politicians who had brought about the débacle.

It was that awful period in London when the English seemed blind and deaf, when there were still people in the government like Lord Londonderry, who believed "the Germans were really a very nice people, so much like ourselves." The atmosphere became unendurable and at last one night when I found myself making an angry speech in the Savoy Grill, I decided it was time to clear out. I loved England. I had had many happy adventures there. My impulse was good. Like so many other good friends of England, I wanted to warn them. It seemed to me that all but a few of the English—Vansittart, Eden, Duff-Cooper, Rebecca West Churchill and in general the "French faction"—were like spoiled, stupid children.

The next morning I called the United States Lines and managed to get a cabin aboard the "Manhattan," feeling in my heart that I would never again find the England I was leaving. I went alone to Paddington and just as the train was about to pull out I saw Mito Djordjadze, tall and dark and Georgian, running toward me. He was carrying a puppy, quite a big puppy, in his arms. As he came near I saw that the puppy was a boxer, not a golden one like Rex, but a dark, brindle with a black face and one white paw.

Breathless, he said, "I've brought you a wife for Rex!" And I remembered the promise of Mito and his wife Audrey that they would one day give us a mate for Rex.

I took the puppy into my arms and realized that she was thin and ill. "It's the best I could find" said Mito. "You went off in such a hurry. She's just gotten over distemper. The kennel people say she'll be all right."

The train started and I jumped aboard with a puppy I hadn't expected.

There was trouble about keeping her in my compartment and I went to the luggage van and sat there all the way to Southampton. If ever a dog could not be left alone, it was that puppy. She was ugly and thin and ill but in her eyes there was something which I think I recognized even then as humor and gaiety and independence.

Aboard the ship she had to stay in the kennel on the top deck with the other dogs, but I spent nearly the whole of every day with her and saw to it that she got warm milk and brandy. Most of the time she lay listlessly in my lap, but she attracted an admirer who evidently liked frail ladies. He was the biggest, ugliest English bulldog I have ever seen. His name was Harry and he had a heart as big as he was ugly. He was being shipped to a kennel in America and he had no master but he soon took care of that by adopting me and the pup. All the way across the Atlantic he insisted on lying on the bench beside us with his head on my knee. By the time we reached New York I would have bought him if I had known whom he belonged to. We said goodby to Harry and he was put back in his box to be shipped someplace in New England. I have never seen him since but he was one of the most charming, if sloppy, personalities I have ever met.

It seemed obvious that the wife of Rex should be called Regina and so that became the puppy's name. She was still very sick while I stayed in New York and most of the time she spent on a sofa in my mother's flat, languishing like a mid-Victorian invalid.

Like many a woman who is never strong until she has a baby, Regina was sickly until she had her first litter of six pups at Malabar. From that moment on she was strong as an ox. Since then she has had twenty more pups and is many times a grandmother, but even today none of her grandchildren are gayer or stronger. She has always been a good mother, but a stern disciplinarian, with no Oedipus nonsense about her. With each litter there comes a time when she feels they should go on their own and out they go. She will have no more of them but in an odd fashion she continues to discipline them. Even though Prince, her eldest son, is only about eighteen months younger than herself, she will take no nonsense from him or from any of her other children or grandchildren. She is affectionate but independent and hardy, and often enough when all the other boxers are inside in winter, Gina stays out sunning herself by the greenhouse with the cocker spaniels.

Rex and Regina became the founders of a long line of boxers at Malabar. Three of their sons—Prince and Baby and Smoky—still are there and will be till they die. Their brothers and sisters are scattered all over the country, with owners who feel about boxers as we do, that there is something special about a boxer which makes them different from other dogs. There is a kind of fraternity among people who own boxers. You can speak to them on the street although they are total strangers and stand there indefinitely talking about the virtues and personality of the breed. I have in my life owned fifty or sixty dogs of many breeds and I have always made friends with dogs, even with poor savage Marquise, but the boxer is different from them all. He is stubborn and gay and comical. He may be devoted to you but never in a worshipful way. He knows your faults and accepts them. He is not a pet. He is a companion and friend and equal. Wherever I go on Malabar, five or six boxers go

with me. They are as companionable as any friend. They race off after rabbits, vainly, for they have little nose and none too good sight, but their sense of hearing is fantastically acute, which accounts, I suppose, for the fact that they have always been watch and police dogs. Their origin as a breed is somewhat obscure. But there is one story that they are one of the oldest breeds in the world, coming originally from north China where they guarded caravans and compounds, and indeed it may be true for the ancient stone dogs one sees in China guarding the entrance to compounds are very like the modern boxer.

They are good farm dogs, for they do not go off hunting as hunting dogs will do, nor run across country chasing cattle or killing sheep like the terriers. Like Rikky the mongoose, they are happiest when they are with people and rarely go a hundred yards from the house unless someone goes with them. To them, the prospect of going for a long walk across the fields is as exciting as going for a walk can be to a dog bred in a city apartment.

The children of Rex and Regina each has his own personality. Prince, the eldest son, never leaves me by more than three feet from the time I rise in the morning until I go to bed at night. There is nothing groveling or worshipful in his devotion. It is simply that I am his best friend. He understands now that when I go away, it's not for good and that I always come back, but the mere sight of "store" clothes brought out of the cupboard throws him into a depression. From that moment on, until I leave the house, he sits at a distance from me watching every move with a sad reproachful eye. Even when I get into the car to go to the station, he will not come near me. He is a great worrier and always filled with anxiety.

His brother, Baby, was in a sense an orphan, for he was the only one to survive from a litter born prematurely to Regina after she threw herself gaily into the middle of a dogfight. He was brought up on a bottle by Venetia Wills, who as an *evacuée* from bombed England was staying with us, and he has his own special character. He is both a born farmer and a clown with a great attachment for the horses. All his days are spent out-of-doors on the farm, mostly with Charley Martin, who runs the vegetable garden and does a bit of everything. He will spend a whole day walking up and down a field beside the horses when they are plowing, perfectly happy. He has never been popular with the other dogs, chiefly I think, because they recognize him as a show-off and a ham actor. He drinks Coca-Cola from a bottle and water straight from the tap, lapping the water as it falls. To him, a wheelbarrow was meant to ride in and if he is with you, working in the garden, he will jump into the wheelbarrow the moment it is empty and stay there until he is wheeled up to the house. But his best trick is his high-diving from the high platform into the pond just below the Big House. With a running start he will go off the platform at a height of twelve feet and leap twenty feet into the water for a stick. Despite the fact that boxers are not water dogs and Rex and Prince had to be driven outside on a rainy day to step high and with distaste in the wet grass, Regina, Baby and Smoky take to the water like retrievers. I think this is so because they grew up with two cocker spaniels and a golden retriever who, even in zero weather, would break the ice of the pond, go for a swim and then come out and roll in the snow. During the hot days of midsummer, Regina will sit in the pond a whole morning with only her head above the water, looking for all the world like a rather ugly hippo.

No one on the farm ever bothered to teach them any tricks, for there was never time. Baby invented all his bag of tricks, begging as a born show-off to perform the high-diving stunt whenever visitors arrived. He is the only one of the dogs who holds long conversations with you, answering each question with different intonations, a trick which infuriates the others, notably his older brother, Prince, who, since the death of Rex, has taken over as boss of the farm dogs. He keeps them in order, aided by Regina, who has never given up disciplining them, even Prince; and as Baby grew older, he turned out to be the only rebel against authority. The rebellion leads occasionally to fights such as took place more and more frequently as Rex grew old and ill and was no longer able to assert his authority. A fight among four or five big boxers is a terrifying spectacle to behold for one not used to them and unable to assert his authority. I have scars on one wrist and one ankle where Prince, in the midst of a fight, got hold of me instead of Rex or Baby. His shame when he made the discovery that it was I he was biting, was moving to behold.

In the evenings all four, Regina, Prince, Baby, Smoky and a newcomer called Folly, come into the house and now and then stage a fight which wreaks havoc. As the source of most such rows is,usually the question of who can sit or lie nearest to me, the fights are likely to take place under a card table. As a fight between two becomes a general brawl, cards, drinks, score pads and all are likely to go flying into the air. Now and then a fight occurs under a table with *bibelots* or a vase of flowers and the result is the same. On occasion the breakage has been expensive. The most celebrated fight of all took place under the dining room table during a children's party with twelve kids present. Before it was stopped, two chairs were broken and there was milk, cake and ice cream on floor, walls and ceiling.

Big dogs, indeed, have no place in the life of demon housekeepers. Fortunately, all our household prefers dogs to an immaculate house. The five boxers sleep in my room which serves as office, bedroom and workroom and fortunately is on the ground floor and has many windows and two outside doors leading into the garden. Prince sleeps on the foot of my bed, Folly in a dog bed at the foot, Gina on the sofa and Smoky and Baby on chairs. Although Baby has long outgrown the chair he chose as a puppy and hangs over both ends, nothing will induce him to give it up. Each place belongs to a different boxer and if one attempts to take over the property of another, the fight is on. Now and then, other members of the family complain that my room is a little "high" and Nanny attacks it with Lysol, taking over usually while I am away and cannot protest. I am afraid that I prefer the smell of dogs to the smell of Lysol.

Last summer a guest observed, "I can't believe this house is new. It looks so old and well-worn. Nothing looks new!" to which George replied, "Dogs and children take care of that. It looked like an old house six months after it was built."

Prince has his own intelligent tricks. The doors of the house do not have knobs but French door handles and almost at once Prince learned that he could open any door at will by turning the handle. It is impossible to shut them in any room. The moment a breakfast tray goes into the bedroom of any guest, it is a sign to the dogs that the day has begun and a procession led by Prince who opens the door, visits the guest. They learned long ago that few guests resist the impulse of feeding them bits of toast and bacon.

Prince not only opens all the doors inside the house but learned long ago to open not only the outside doors, which do not have handles but thumb catches, and to hold back the screen door while he opens the main one. He knows too how to open the doors of an automobile and more than once visitors have looked out of the window on cold days to observe five or six boxers sitting in their car with the doors all *closed*. For Prince learned too that closing the door behind him prevents cold air and drafts from entering the car.

But the ancient Ford station wagon is their delight, for they know that when any of us get into it, it means that we are off to some remote part of the farm where the other cars cannot go because of the mud or the roughness of the roads and fields. In summer in the early morning they will go and sit in it to prevent its getting away without them. Once it arrives in a big field they are all out at once to run and hunt. They are not clever hunters and the only thing they ever catch is ground hogs which abound over the whole of the thousand acres. They know every ground hog hole on the whole of the farm and long ago learned the trick of going straight to the lair as quickly as possible to get between the ground hog who may be out feeding and his safe, underground lair. Each one of the dogs, as a pup, had a good mauling by a big ground hog and the lesson has never been forgotten. Rex, the German, nearly lost an eye in his first encounter. A big, old ground hog can weigh as much as forty pounds and will fight and claw like a wildcat, but the boxers learned long ago that the trick is to get their opponent behind the head and break his neck. It is all over in a second.

Although killing anything gives me no pleasure, the dogs do serve as a check on the ground hogs which otherwise might dig us out of house and home, for with lynx and wildcat and wolf gone from our country there is no longer any natural check on the ground hog. No fox dares attack them. Their digging can start bad gullies and they can destroy a whole young orchard in a week or two. There are still plenty of them left to dig holes and make shelters not only for themselves but for rabbits and other game. A female raccoon, hard put to it for a lair in which to have her young, will share a ground hog hole with the owner.

There are plenty of dogs on the place, for Bob has a female boxer called Kitchey, and Kenneth one called Susie Parkington who also knows the Ford station wagon trick. If Susie sees or hears the station wagon, she will come across country any distance to join in the picnic.

There is no more sociable dog than a boxer and there is nothing they like so much as visitors or a party. As they come rushing out they are likely to scare people to death, since they are ferocious in appearance, but

they are in their hearts all amiability. In that respect they are ideal watch-dogs on a place like Malabar where there are hundreds of visitors a year. No one can come near any building without the boxers knowing it and setting up an uproar, but biting is not a part of their natures. The golden retrievers were different for nothing could persuade them that a part of their duties as watchdogs was not to *bite* and on occasion they went beyond bluff and took pieces out of Charley Kimmel, the game warden, whom they knew well, and out of two or three other visitors. In the end we gave them away, although they are the most beautiful of dogs.

The boxers, used to square dances and picnics and meetings, learned long ago that a group of automobiles meant a party and a party meant that they were going to have a feast of hot dogs, of steak bones, of pie and cake and doughnuts. No party ever had more welcoming hosts than the boxers and the two cockers on the occasion of farm festivities. Usually the next day meant indigestion for them and the consumption of great quantities of grass and zinnia leaves, which boxers seem to regard as a cure for indigestion. But apparently the party is always worth the indigestion.

I cannot write of the dogs at Malabar without mentioning the two small female cockers, Patsy and Dusky, who came to the farm as pets of the children. Dusky is all black and very feminine and very sporting. When the dog party goes out in the old station wagon, Dusky and Patsy would go off hunting, *really* hunting and not just rushing about rather aimlessly like the boxers, deep into the woods to come back all wet or covered with burrs hours later. They were inseparable companions who lived out-of-doors and slept in the potting shed; silky, affectionate and charming.

Of the two, Patsy had the eccentric character. She was a very small cocker, black with tan-colored spots and big brown eyes with a slight squint which gave her a hopelessly comic expression. Despite her small size and good nature she took no pushing around from the big boxers. Although in play they sometimes sent her rolling end on end, she always came back making hideous snarls and growls. In a way, her life was a tragedy of frustration, for her one overwhelming desire was puppies. With her first litter, she caught pneumonia and all of them died. Shortly after that she ran under a car and suffered a broken pelvis. It mended badly and after that there was danger in her having puppies and we did not breed her, but twice each year she had an hysterical pregnancy even to producing milk. Whenever Gina had puppies which was about once a year, Patsy acted as nursemaid. Apparently she and Gina had some agreement for although Gina would allow no other dog, male or female, to come near her puppies, she made no objection to Patsy sitting beside the straw-filled box in which they lay or even to sharing the box with her and her puppies. Patsy would never leave the vicinity of the box, save to eat, and when Gina went out she would get into the box with the puppies and even attempt to feed them. I think she got satisfaction out of all this and she helped Gina to bring up four litters of pups, which before they were three months old, were twice as tall as Patsy herself.

This last summer she showed all the signs of hysterical pregnancy again and none of us, accustomed to the phenomenon, took any notice. But this time the pregnancy was real. Somehow she had been bred, probably to one of the big boxers. The puppies were born dead and Patsy died soon after, happy, I think, in the illusion that again, after years, she had produced a family. Anne wrote a poem about her—

It matters little to the world
That one small dog
No longer trots these dusty lanes.
The canny eyes, untidy hair
Brown dots for eyebrows
Have faded in the dark, unknown of death

Much like a pebble in a dark smooth pool.
Still she is here, trotting busily
The steep fern-covered hills
So long as we who were her friends, remain on earth.

To all the family at Malabar, animals somehow are not animals and are not treated as such. I believe that all animals have a dignity, an integrity and an entity of their own and men get back from them in understanding and loyalty and quality of understanding exactly what they give.

For us, Tex is not simply a beautiful mare, but a queen who rules all the other horses, imperious and conscious of her beauty. She leads and dominates them all, even the geldings. It is useless to attempt rounding up the other horses unless Tex is first under control; and being clever she is hard to catch in an open field. She knows the tricks of luring her with an apple or an ear of corn and can take it from your hand quickly, too quickly to permit you to seize her halter. She understands when someone who does not know how to ride climbs on her back. She will shy, dramatize every rabbit or blowing leaf and pretend to run away with her rider, like as not to leave him by the side of a lane or in the middle of a field, quietly, without hurting him. With an experienced rider on her back she is a different animal. She will show off her five gaits and even attempt in the middle of a pasture to go into circles and figure eights, as if she were back again in the show ring. In the field, she takes care of the blind Percheron mare, Sylvia, never going far from her and whickering to lead her across the creek or up a steep bank. When Sylvia feels herself lost, she will whicker frantically until Tex answers.

And there is Tony, Hope's big pony, who is a clown and shakes hands and does other tricks, one of which is to steal up behind you in the pasture and butt you in the middle of the back with his head. He is neither bad-tempered nor bad-mannered but frolicsome, and it is always well to keep out of the way of his heels if you are on another horse. He once put me out of the running for three days with a side kick delivered on a gallop as he passed Tex and me. Only the steel stirrup saved me a broken leg.

And Red, who is twenty-six years old, takes care of the younger children. Any babe can ride him. He even glances back occasionally, I think, to see that the small ones are firmly in the saddle.

None of the children has ever been afraid of animals; all have always accepted them as companions and equals. In the beginning at Malabar when they set out on a walk they were always followed by four or five dogs, four goats, a tomcat and a pet lamb.

The goats came into our lives when Kate Tobias, an old friend and schoolmate of mine who kept a grocery store, a filling station and a stable of goats on the Little Washington Road brought a tiny kid as an Easter gift to the children. Nothing in life is, I think, prettier or more touching than a kid, with its deer-like ears and big brown eyes. The first kid called for a companion and we ended up with four. As they grew up as part of the family, they wanted no part of life in the fields. They wanted to live *in* the house and their favorite resting spot was the porch swing where all four would lie, gently chewing, to watch what went on in the lane and barnyard. Twice they managed somehow to get into the house and were found on the bed of a ground-floor bedroom. Quickly they learned to leap to the hood of a car and thence on the roof and after that no car ever stopped at the door without all four springing to the roof where they lay peacefully chewing until driven down.

It was impossible to shut them up and we tried vainly to induce them at last to live in the pasture with the sheep, but they would have none of that and always they would beat me home from the pasture and be there on the porch quietly rocking in the swing to greet me. They would simply walk up the brace of a fence or lie on their sides and squeeze *under* the fence. One of their favorite pranks was to walk across the top of the hotbeds, breaking glass as they went. In the end they became intolerable. It seemed that at times they actually persecuted us with a special goatish intelligence. The decision to save ourselves from them came at last when one summer day they got inside the car of a visitor and were found placidly chewing on a scattered packet of neatly stamped and addressed envelopes. To save ourselves from further persecution, we found homes for them in the suburbs of Cleveland among the Italians who undoubtedly knew better how to cope with goats than we did.

But I am not sorry for the experience of the goats, for it gave me an insight into all the legends concerning goats that have been a part of human history and mythology since the beginning of time. Their intelligence is of a different quality from that of all other animals, almost human and a little devilish. The ancient wisdom in the eyes of a billy goat can shame you. And at courting time they become very nearly human. A billy goat prancing on his hind legs, tilting his head from one side to another, making sounds that have strange half-human intimations makes the legends of fauns and satyrs very real and very near. No animal is so amorous or so demonstrative as a he-goat and none more human. Hector, the big Nubian he-goat, was, when courting, the perfect picture of a lecherous bearded old reprobate. He seemed strangely out of place among the soft green hills and woods of Ohio and indeed, almost obscene. He belonged, I think, among the gray rocks of Italy or Greece, prancing and strutting amorously among the olive trees.

And there has been a long line of orphan lambs brought up on bottles as pets by Hope and Ellen. No animal is more touching than a young lamb and no animal more stupid and lacking in charm than a grown sheep, and so as they grew up, interest lagged and they were returned to the flock, or at least we hoped they were. But they wanted no part of the sheep. They had been brought up among people and *dogs* and *horses* and despised the other sheep. When you crossed a pasture, they would leave the sheep and join you and the dogs, and always they kept returning to the Big House. As Ellen once said, "They don't know they're sheep. We should get a mirror and let them look at themselves."

Certainly one or two of them thought they were dogs. They ran about with the dogs and even chased cars with the dogs and nothing startled a stranger so much as to see a good-sized lamb frolicking with the boxers and sometimes in the middle of a dogfight. It was only when breeding season came that the lambs finally gave in to the fact that they were, after all, sheep and we were rid of them.

And there were the karakul lambs given Ellen by Grove and Esther Patterson—strange-looking beasts more goat than sheep, who would rather eat weeds and briars than good bluegrass and clover. They came as tiny black lambs, with silky tightly curled coats and long drooping black ears. They grew into a ram and a ewe, still black until they were shorn for the first time. The ram grew great curling horns and like most of the animals simply acquired a name which established itself. He had a long aquiline nose which with his thick black coat and his melancholy dark eyes and skinny legs begged for the name of Haile Selassie. As George observed, "All he needed was a royal umbrella to look exactly like the Lion of Judah."

Haile had a playful disposition and a sense of humor and usually frightened visitors by backing off, rearing on his hind legs and springing toward them. It was rare that he actually butted in earnest and even then he never did much harm, for he was mostly wool and despite his size, weighed only about half the weight of a Dorset ram. He insisted on going on walks along with the dogs and occasionally had to be lifted over tight wire fences. He became quite accustomed to the hoisting operations and at sight of a fence would place himself alongside it and wait to be lifted over. More than once he has butted visitors gently through the front door, on one occasion no less a person than Vladimir Popoff, animal breeding expert from the Soviet government, who had come to the farm to see our blue-roan steers. After having been knocked down by the welcome of an array of affectionate boxers accompanied by two lambs, and butted into the front hall by a big karakul ram, he observed, grinning, in reply to my apologies, "But this is the first time I've felt really at home in America."

Alas! Haile developed a nasty habit of *hooking* you with his big horns. Since he usually caught you on the shins, it was a painful habit and we had to shut him up in the paddock overlooking the courtyard at the Big House. Here he spent his days on a cattle-loading platform watching the traffic of the farm and conversing with passersby in deep-voiced bleats. When there were no strangers around I would let him out and play a sort of bullfighting game with him, letting him charge and catching him by the horns just as he got to me. He would play this game over and over with zest until I was worn out.

Now he is living with his mate Snowball and a pet Dorset-Delaine cross ewe who was brought up as a pet. If you cross their pasture, they will follow you and Haile will begin his old tricks. You can call him from any distance within hearing range and he will raise his head, bleat and come running toward you.

Most of the animals named themselves. That too was the case of the wild tom turkey we bought to cross with the bronze hens in the hope that the crossed offspring would develop more sense than pure-bred turkeys about taking care of their young. The experiment was a failure for the cross-bred hens proved just as idiotic as the original bronze stock. But the tom turkey, like the goats and lambs and Haile Selassie, developed such an attachment to the family that he wanted to live in the house, and like them, spent a good part of his time following you about or looking in the windows. He was and is anything but a wild turkey. In summer, you cannot be rid of him, and his strutting and gossiping and gabbling sometimes drowns out conversation. He is a handsome animal and knows it. For hours he will strut beside a car, regarding his beauty in the enameled reflection. He is a bird of violent and implacable antipathies, especially for anyone who has ever teased him. He has a hate for Bob and will give him no peace when he is at the Big House. He will follow him about, uttering high-pitched peevish cries, and attack him with wings and feet and beak at every opportunity. He will chase Bob's car all the way down the lane as far as the road. He has a similar hate for the driver of the bakery truck and for two of the boys who work on the farm in summer. And he carries on a feud with Baby, the clown boxer, who rushes at him and strong-arms him without ever doing him more harm than knocking him over. He acquired the name of Gilbert in honor of a fat acquaintance of the family who never walks but struts pompously.

No chapter would be complete which overlooked Donald. I found him in the barnyard, when he was only a day old, wandering bravely about, a tiny ball of fluff. He seemed to be without father or mother and to this day I have never discovered where he came from. I took him up to the Big House and gave him to Ellen who was in bed with a cold. She put him in a shoe box and kept him on the bed with her day and night. When she got up he continued for a time to live in the nursery where he ran across the floor chasing moths brought in for him. Usually the chase ended in a wild and comical skid on the slippery linoleum. He would climb all over you as you lay on the floor and grip you firmly by the ear in his soft bill.

As he grew older Ellen kept him outside where he followed her about wherever she went, and although that summer she tended and brought up five hundred ducks, he never showed any desire to leave her and join his own kind. As George said, "The trouble with the animals on this farm is that they all think they are people."

Then Donald went through the ugly duckling stage between down and feathers and presently grew a coat of feathers which gave some clue to his parents. He came out speckled black and white which led us to believe that his father must have been a half-wild Muscovy duck which flew in one day out of nowhere to join the Pekings on the lower pond. He still chose to live at the Big House, ignoring the other ducks and living peacefully with the dogs, the pet lambs, Gilbert, the tom turkey, and the fighting chickens. He never showed any fear of the dogs and would boldly waddle across them as they lay on the grass. It was not until the wild ducks, coming south for the winter, descended on the ponds that he gave any sign of knowing that he was a duck. Tentatively, at first, he would join them for a swim, which led us to suspect that somewhere in his ancestry there was a mallard or two. But for a long time he returned every evening to the garage to sleep alongside Dusky and Patsy, the cockers. And then one day he went off the pond for good and turned out in the classical manner to be a she but "he" would always come when you went to the pond and called to "him." He would leave the other ducks and follow you along the border of the pond quacking in a friendly fashion.

And there was the spotted Poland China sow who became known spontaneously as Rachel because, when seen from the rear while walking, she resembled nothing as much as a female resident of the Bronx walking down Broadway in a mink coat.

I think I love as much as any animal I have ever known, the big Angus bull—Blondy. He has an extraordinary sleek, inky, black beauty, with huge muscles and a giant arched neck, with great black intelligent eyes and a friendly disposition. He lives mostly on the range with a harem of thirty shorthorn cows, but you can go up to him any time in the field and scratch his ears without his ever showing either nervousness or anger. Now and then we have attempted to shut him up at certain seasons, but always very quietly and thoroughly he used his great shoulders and head to demolish fence or pen or even the side of the barn, and with great dignity and calm he set out across country over fences to rejoin his harem. Once we attempted to keep him in by using an electric fence. The shock awed him into submission and temporary obedience, but Harry turned off the current at night for reasons of economy. While the current was on, the transformer made a slow ticking noise and Blondy, after three or four days, came to understand that when the current was turned off the ticking stopped and the wire was harmless. The first morning after the observation, he was gone again back to the high pasture with his harem.

And so the stories of the animals at Malabar go on and on. It is a friendly place because of the animals, who to me and indeed to all of us, each has a personality. Fortunately Harry and Bob and Kenneth and Charley feel much the same way and I think the animals know it. They are good men who understand livestock and could not sleep if any animal were sick or cold or without feed or water. And the children are growing up with a feeling of sympathy and responsibility toward all dumb beasts, a feeling which can bring great richness in life and great understanding of things which others, who do not know that sympathy and responsibility, never understand or fathom. And the odd thing is, that the feeling extends to the animals themselves which live together—dogs, barn cats, and lambs in peace. There are times in my own relations with the farm animals when I am tempted to accept the beliefs of the Hindu and the Jain concerning reincarnation and the sacredness of all life. In any case, I know how much poorer life would be without the animals and their trust.

RECAPITULATION

"In order to subdue Nature, you must first understand her."
—Francis Bacon

IN THE WHOLE STORY OF THE PLEASANT VALLEY ADVENTURE, none of us engaged were inexperienced innocents. All of us in one way or another had come from the soil itself or at an early period discovered that it was in our blood. Only one of us, Bill, had turned toward the high wages and the precarious existence of the industrial worker, and Bill's story is not yet finished. He found very shortly that by the time he had paid rent and union dues and assessments and retail city prices for the food he ate and contributed to Community Funds and went three or four times a week with his wife to picture shows, he had less left in his pocket than when working for lower wages on a farm or than operating his own farm, however small. And just ahead of him always lay the threat of a depression and unemployment, balanced only by the hope of a dole in the form of social security payments or work on the humiliating level of W.P.A. wages.

And there were other elements, just as important, perhaps more so, which had nothing to do with money. Before a month had passed Bill returned to see us in "store clothes" and new foot-pinching shoes. When you asked him how he liked standing all day in one spot on an assembly line, fitting gadgets together, his first comment was "my feet hurt!"

There were other comments and remarks, some quite frank, others sheepishly veiled or implied. Gone were the days when he could ride a tractor, free as air, round and round the long slopes of the Valley hill-sides. Gone were the times when if the day was hot, he could stop the tractor and walk a few yards and go swimming naked as Adam in the cool deep holes of Switzer's Creek and the Clear Fork. Gone were the moonlight skating parties with the Farm Bureau young people or the Youth Cooperative. Gone too were the times when, if thirsty, you could walk into the springhouse and quench your thirst with a glass of cold Guernsey milk or the icy, *living* water which flowed out of the pink sandstone rock. In the city flat where he lived there was no cellar at all but only a cubicle for a kitchen with cupboards and icebox which would hold little more than a day's supply of food in advance. There were no golden hams or flitches of bacon, no rows of pickles and home-canned peaches and wild blackberries, no crocks of butter or buttermilk, no heaps of Hubbard squash and country cabbage. It was a slut's kitchen, where everything came out of cans if you didn't go out to the sleazy restaurant on the corner.

There was no big fireplace or red-hot stove but only a hole in the floor out of which came dead, dusty hot air and when you wakened in the morning, you did not look out over the Valley, smiling and green, with the little creek wandering through the lush bluegrass pastures where the Guernseys fed. You looked out of a dirty window into a street where no matter how often you painted the houses nor how varied the colors you used, they were always soot-colored. And you wakened in a world blanketed by soot and smoke where even at noonday the sunlight was pale and sickly. Bill hadn't any children yet, but when he does have them, that is the world in which they will grow up into men and women.

He told us it wasn't too bad because his shift came off at four o'clock and he could always go down to his father's farm and work until sundown—when he had sufficient gasoline.

I think it won't be long until Bill is back with us, or until he has a little place of his own outside in the country where there will be sunlight and a garden and some chickens and a cow and a couple of hogs to give him a sense of the only *genuine* security that exists on this earth. And Bill, day by day, is not dealing with life. He stands on a cement floor and all day fits one dead object upon another, over and over again, a thousand times a day, and while he is doing this he himself is slowly dying inside. When his day is finished and he is short of gasoline, he has no place to go but the flat with the sooty windows or the corner poolroom or the saloon or a picture show. He is bored with a deadly boredom which no amount of silk shirts or expensive wrist watches can ever assuage. And each day he grows a little more surly, a little more certain that somebody should pay him a lot more than he is worth for doing something he does not want to do.

Each time he comes to see us, you see his dull eyes brighten when they light on a young heifer or a newborn lamb, or a field of tall dark-green corn growing so fast that you can hear its joints crack. He will be back with us someday or he will get himself a little piece of land and a little house to save his own soul, or he will turn into a brutalized, sullen man with children who will have small chance of growing up into healthy, useful citizens, because you cannot do violence to Nature herself without paying for it as an individual or as a nation. The monstrous, ugly cities which have grown up about steel or rubber or automobiles, are in themselves violations of Nature which can only produce monstrosities in the form of individuals, of social unrest, of wretched insecurity, of delinquency and vice and class feeling. They are in themselves as perverse and murderous as Jack the Ripper.

There are signs, many of them, that both organized labor and industry are recognizing the horror of crowded communities with factory piled upon factory and slum upon slum. It may be that the cumulative effect of these smoky compounds of "civilization" will in turn be responsible for a revolution. The latest census showed that many cities were losing rather than gaining population. Each year the little ring of small houses and gardens and little farms surrounding the cities which are not too overgrown, becomes a little wider. Those rings are actual rings of security, both for the individual and for the state—a genuine security not founded upon taxes and bureaucracy and doles but upon Nature and common sense and fact; breeding dignity and independence in man rather than surliness and bitter resentment. They offer not only economic but, what is of greater importance, spiritual security. They are a safeguard against the most ugly of things—a proletariat—which is no more than a mechanized agglomeration of robot citizens without dignity or independence.

Industry itself moves toward decentralization. There is no real reason beyond transportation and raw materials why all of our industry should huddle together in one region. And certainly there is no reason why in one region, they should be crowded together in an area of a few

square miles. Such developments grew, not out of human intelligence, but out of human greed, opportunism and stupidity.

During the years of our black depression, Europeans frequently asked me, "Why in a country as rich as America should economic depressions be more desperate than anywhere else in the world?" Why, indeed? Why should our latest depression have been accompanied by suffering and death as great as that in a nation like Germany which had real and desperate reasons for social and economic depression. Germany, at the period of world depression, was an overpopulated nation crowded within the borders of a country of small and poor resources. She was a nation defeated in war, with a currency inflated to the point of invisibility. She suffered from grave civil disorders and wholesale individual bankruptcy, revolution and fundamental shortages of food and raw materials.

The United States, on the other hand, was an underpopulated nation with vast natural resources, suffering no inflation, paying at that time only moderate taxes, with enormous internal and foreign markets, with two-thirds of the gold of the world to support her currency. History, I think would say that in the abstract there were no reasons for the colossal collapse of a great nation's economy, that none of the fundamental causes were present. Yet we had more than twenty million unemployed, with soup lines in which thousands stood in line waiting to be fed by charity, war veterans and old women selling apples on street corners, death by starvation, a colossal dole in the form of W.P.A.—a whole nation reduced at last through heavy taxation to a point at which it had begun to devour its own vitals.

The causes were many and complex. To set them all down intelligently would require a very long book. Most of them, I believe, were "artificial" causes and the product of our national characteristics and folly—our gambling instinct, our lack of all sense of thrift, our national habit of always doing business upon credit or borrowed money, the vicious improvidence engendered by the installment plan, the wastefulness, ignorance and greed of our reckless attack upon our own natural resources, the sickness of our agriculture, the inflation of stock market values through gambling, the over-centralization in New York of our whole financial system, overexpansion and over-production without adequately developed distribution; and most of all a growing industrial population of millions crowded into hideous cities where high costs devour the value of every dollar. No nation in history has ever presented a picture of greater folly and confusion than this one during the period when in cities like Omaha and Kansas City people stood in soup lines or occasionally died of starvation while farmers a mile or two beyond the city limits killed and burned pigs and used corn and wheat as fuel in their stoves!

The "artificial" causes of our shameful economic collapse are indeed many and diverse, but below and beyond them all, one cold and fundamental fact is apparent and that is the vast individual improvidence and the utter economic insecurity of the great majority of the population of the United States, especially of the so-called working classes, the white-collar workers and a large proportion of those engaged in agriculture.

Any nation, and particularly a democracy, is only as strong as the sum total of the individual citizens who comprise it. A paternal government, a benevolent tyranny, a Fascist government may manage to create for a time the illusion that this is not true but in the end none of them can escape the inevitable results of a miserable or insecure population. A democracy in the face of collapse is by the very fact of its virtues, naked and helpless in the face of that collapse. Many of the most bitterly resented attempts of the New Deal to regulate and check the private affairs of the citizens of a democracy grew out of the unconscious desire of many otherwise worthy men to impose Fascist or socialistic controls in order to cover up the helplessness of democracy in the face of economic disaster. None of these controls, even the soundest, were able to conceal the fact that when the whole structure began to totter, there was and is no way of cushioning or checking collapse so long as the great majority of the population is without resources or security. Rome collapsed, despite every desperate measure, when she reached the point where two-thirds of her population became indigent and took to huddling in the cities to live off state charity.

I know of no better way to illustrate the point I wish to make than to take the relative positions of the American workman and the French workman when faced by unemployment and economic crisis. I have divided my life about equally between America on the one side and Europe and Asia on the other, and with the case of the French workingman, I was closely associated over a long period of years through a vast French organization known as the Workingmen-Gardeners' Association. It had several million members and the implication of membership was simply that all its men should be industrial workers or white-collar workers or small shopkeepers who possess a piece of land either by long lease or by outright ownership.

Let me take the case of Bosquet, one of the workers who for years was an intimate friend of mine. He had about two acres of fairly good land with a little house on it adjoining our own small farm. He was thrifty. Whatever he had, he *owned*—his little house, his tiny piece of land, his furniture, his old-fashioned gramophone and the records which went with it. He was a machine-tool worker, employed by the French subsidiary of a big American manufacturing company and worked at Le Bourget, about twenty-five miles away. Nothing could have induced him to live *in* the industrial suburb of Le Bourget or in one of the tenements built for workingmen further in toward Paris itself. Sometimes he went to work in the bus. Sometimes he drove to work in an eight year-old car which he rebuilt from time to time. As wages, he received about a third of the amount paid to an American worker with the corresponding kind of a job.

The little house was built largely by himself with the aid of his friends. On the two acres of land, Bosquet, with the help of his wife and three children, raised virtually everything the whole family ate during a year. On the two acres he grew every sort of vegetable including such luxuries as artichokes and strawberries. He kept about twenty-five chickens, a little flock of ducks, a pig and two she-goats. It was a small compact world, operating in close alliance with Nature herself. The livestock made manure which went back into the soil, along with all sorts of rubbish dumped into a constantly ripening compost heap, to make in turn the vegetables which fed the family and partially fed the livestock. The she-goats largely found their forage along the roadside.

Not far away was the state forest from which the Bosquet family got their fuel free. Under French law the citizen has a right to all timber lying on the ground in a state forest and a couple of times a year the Bosquet family had a picnic over week ends in the forest and came home with the wood supply to last them several months. At the end of the garden near the clear little river which fed the cress beds, Bosquet built himself a summerhouse and there on summer evenings he would sit with his friends by the hour, talking politics, drinking wine and singing to the accompaniment of his concertina. His children had thousands of acres of meadow, forest and river in which to enjoy themselves and grew up into healthy good citizens, with well-balanced, unjaundiced minds. Bosquet was perhaps the happiest man I ever knew, largely I think, because he was living as Nature meant a man to live and because he knew his little world was secure because it was built upon fundamentals.

Twice during the time I lived next to Bosquet he experienced long periods of unemployment, when hard times shut down the factories. One period was particularly grave for it came at the time when inflation drove the franc to an exchange value of over fifty to the dollar. In neither of these periods did Bosquet or his family suffer. He simply "dug in" and held on. His family never went hungry; on the contrary they were better nourished than many a city dweller who had not lost his job. Nobody took away his jalopy or his furniture because he hadn't paid for them yet. They were his property. Bosquet never went on relief. He didn't take a W.P.A. job because there wasn't any such thing or any real need for it; in France there were too many other workmen who shared Bosquet's dug-in, natural security. At a time when five or six thousand unemployed stood in line at soup kitchens in our bigger cities, I never saw in the industrial suburbs of Paris, through which I passed several times each week, more than fifty or seventy-five unemployed in any one line. They represented, I suppose, about the proportion of naturally shiftless and indigent individuals to the whole of the French working class.

And now let us take the case of an average American individual worker. Call him Joe Smith. He lives in a rented house. He has his automobile, his radio, his washing machine on the installment plan. His children grow up as often as not in a slum neighborhood and go to crowded inefficient schools where they are subjected to low ideals. He buys everything that comes on the table, a large part of it out of cans. A slump happens in his factory because there is over-production or bad distribution and he is laid off. The chances are that he and his family will be in the street on the following day because he has little or nothing laid away. The automobile, the radio, the washing machine go back to the dealer because he cannot keep up payments. He loses the money he has already paid on them and the dealer, who does not want them back (since in a depression there is no longer any market for them), loses what is not paid on them.

Joe's private disaster is one stone dropped into a pond but its ripples are far-reaching. The turned-back automobile for which he has not paid causes a loss to the dealer and manufacturer. Multiply Joe's case a few times and the whole thing begins to pyramid toward disaster. More men out of work means less money spent, less commodities bought. Other factories begin to reduce production and close down and more Joe Smiths are dumped into the street because they have no more security than he had. Dealers close shop because they are no longer able to pay the overhead, for all too often they themselves have no more security than Joe Smith. The economic plague begins to spread like typhus or cholera as the economic structure collapses more and more rapidly. Higher and higher taxes to support Joe and his friends are imposed upon those who through caution or enterprise and thrift still have a job or a little money. Individual enterprise is throttled because income and capital are too heavily taxed or because the risks of investing money at such a calamitous time are too great.

That, it seems to me, is how you can have a terrifying depression in a rich, underpopulated nation in which none of the fundamental causes of an economic depression are present. To be sure, the example used is a highly simplified one, but the elements are present in every industrial center of the United States.

Joe Smith is not alone to blame. Perhaps he is unthrifty, improvident and extravagant. He is so partly because Americans until now have always had a whole rich continent to plunder and have respected and followed the qualities of extravagance, of gambling, of improvidence always in the belief that in a country so rich you could always regain tomorrow what you lost today. And Joe Smith is improvident and extravagant because he is the victim of a high-powered, industrial civilization which has persistently encouraged him to be so. For years it kept persuading him that if he bought more commodities, more would be manufactured and so he would have more work and higher wages. He was told that he could own even a Rolls-Royce on the installment plan and that was agreeable to his ears. He was told that in doing so he was benefiting not only himself but the nation. To turn back now and read some of the advertisements of the insane twenties, one gets the impression that it was the patriotic duty of every citizen to live on the installment plan and to be perpetually in debt. Only one thing was overlooked, and that was the laws of economics which are as immutable as those of mathematics or Nature. An automobile, for example, costs so much in labor and materials. That cost must be covered plus a profit to insure the continuance in operation of the factory which built it. If these costs are not met day by day there comes a day when someone like Joe Smith collapses and then the whole structure presently begins to totter and at last to fall.

The follies of the more moon-eyed of the New Dealers are no more absurd than those of the high-pressure businessmen of the period which eventually gave the more moon-eyed of the New Dealers their opportunity. Both ignored not only the fundamental, immutable laws of economics but both worked to bring about the millennium through methods which deal with superficial theories and conditions, and rarely with fundamental facts. The high-pressure boom-proud businessman managed to put a whole nation into debt and to destroy much of its security by preaching, pressuring and honoring the philosophy of buying more and more whether or not you could pay for it. . . . The moon-eyed theorist sought to correct this by plans which aimed at creating security for the individual through subsidies, pensions, bribes, paid for by constantly increasing taxes upon the very sources of our economic wealth and well-being. And in the pursuit of such theories, they began the structure of a vast and costly bureaucracy which in itself is a kind of economic cancer, devouring as it grows the very theoretical security which was the theoretical justification for its existence. Even today people in the lower income brackets are being taxed so heavily that the margin of savings which might have been used to build up their individual security is being taken from them to support the vast extravagant structure of the very bureaucracy which is supposed to provide them with security.

I have mentioned only the economic aspect of security provided through bureaucracy. The red tape, the irritations, the tyrannies, the sacrifice of human dignity which arise through a security managed by the state is another story which induces a backward trend away from the principles of all democracy and of human dignity.

But we have left Bosquet for a long time playing his concertina in the little summerhouse at the end of the garden. I remarked that Bosquet was perhaps the happiest man I had ever known. I think this is so because of two things. (1) He was as near to knowing security as any human can be. (2) He was really a free man, standing on his own feet in that little world of two acres, infinitely more free than Joe Smith, infinitely more secure, despite Joe's bigger income from wages, despite social security, despite relief and make-work projects.

I heard from Bosquet's family regularly until Germany declared war upon the United States and a steel curtain came down on France, shutting off all communications. Sometimes I had messages brought out of occupied France where Bosquet had his house and garden. Very occasionally I had a scrap of a letter from him. Despite war, despite the collapse of his nation, despite the occupation of his town by enemy troops, despite everything, Bosquet and his family seemed to be doing all right. He was still dug in and sitting tight on his little island of security. His family was still eating. The Germans confiscated food from the big landowners and the big vegetable growers, but Bosquet's security was so inconspicuous that they did not trouble him. He was making enough money to get along under conditions of utter disaster and he was sure that he had a roof over his head and that his family was warm and well fed. When the enemy was driven out of France Bosquet was still there on his two feet, squarely planted, ready to begin the reconstruction, just as his father and his grandfather had been before him. There will be no need to put him on a dole or to compel him to work for humiliating wages.

Bosquet is important as an individual, as individuals must always be in every state, if we are not to sink to the level either of Communist or Fascist regimentation; but Bosquet is of immense importance to his nation and all the other individuals in it because in France there are so many others like him, in the same situation. Three times within seventy-five years France has been the victim of an aggressive war. Three times her soil has been invaded. Twice her capital has been occupied and once the whole of the nation has fallen into the hands of the invader who sought by every vicious means to weaken and destroy her. She has not been destroyed; it is astonishing that she has not been weakened to a far greater degree. Twice, once in 1870 and again in 1918, she came back with astonishing rapidity and vigor. She will survive and be reborn again and it will be so largely because her whole civilization is built upon the rights of the individual and her economy upon the security not of a bureaucratic state but upon that of the individual citizen who in the aggregate makes up the nation and its strength.

Altogether, Bosquet is quite an important fellow.

We, as a nation, have had every possible advantage. We are a rich country, capable, with common sense and intelligent management, of supporting a population two or three times as great. We have never been invaded; our land has never been occupied. It has been easy for us, as a German once said to me, to win every hand in the international game of economics because we always have an ace or two up our sleeves. Yet with all the cards stacked in our favor, I doubt that our economic structure, faced with the violent ultimate strains of war, invasion, dislocation and occupation by any enemy army, would have survived as well as France has survived. Our follies have consistently been those of a reckless son who inherited too much money. That philosophy has for too long—indeed since the beginning of the nation—dominated our thinking and our

social and economic structure. Whether expressed by the businessman in his reckless exploitation of resources or the more extreme New Dealer with his panaceas and economic short cuts, the philosophy has been that of getting rich quick, of getting something for nothing, to the point of killing the goose that laid the golden egg.

The farmer, indeed, has been on the whole no wiser than the rest of us. He has gone ahead all too often operating not as a farmer at all but as a miner, taking everything from the soil and then abandoning it as if it were no more than a coal mine.

And he has been guilty of the same improvident follies as Joe Smith. His land was his capital and persistently he destroyed it. Certainly 80 per cent of our American farms have grown steadily poorer since they were first cleared and put under cultivation. Hundreds of thousands have gone out of circulation altogether to turn from important national economic assets to economic liabilities.

Higher taxes, less profits, less individual ownership, a lowered standard of living, even shortages of certain foods are not necessarily manifestations of temporary conditions owing to war or depression. The level of living has begun to slip downward, and it will continue to do so with an increasing degree of speed, unless somehow we return to fundamentals and cease trying to deceive ourselves with panaceas, with pump priming, with pensions and subsidies and all the apparatus which in any country are not indications of strength but of alarming fundamental weakness. They are of necessity only stopgaps, improvisations, hopes that the crisis can be tided over till "something turns up."

Your European is a good farmer because he has to be. He knows when he is born that the only heritage he will have from his father is the piece of land upon which he is born. He knows in full manhood that the piece of land is his whole capital and that it must support him and his family during his lifetime and as an old man he knows that the piece of land is all he will have to leave as a heritage to his children. Under these conditions you learn to be a good farmer who cherishes his land. Otherwise you are lost, for no one in the end, not even the state itself, is going to pension and support you when the land itself is destroyed. No state can long afford the luxury of creating subsidies drawn by vast taxes from the higher levels of the economic structure in order to support the fundamental industry which should be the foundation of all else.

In our rapid and sometimes reckless industrial development, we as a nation have fallen into much distorted thinking. We have come to think of civilization as entirely a matter of mechanics—of automobiles and cinemas and electric refrigerators and plumbing. As a matter of fact, none of these things has much to do with the advance of civilization, which is a matter of man's education, his spirit and his sense of humanity. Mechanized progress is a matter of comfort and convenience which can have the result of blocking the growth of civilization, distorting it and perhaps destroying it. Certainly there is little evidence that the mechanical things have done much if anything to improve the quality of our statesmen and our thinkers. They have given birth to no new or great philosophies. Only superficially and almost by accident you might say, has the mechanical-industrial development been able to raise the living standards of mankind as a whole, and at the same time it has created a huge urban industrial population, devoid of all economic security, and victim of the alternate booms and depressions which have made up our whole economic history since the Civil War. We have not yet begun to employ this mechanical development to the maximum of the opportunities they so patently provide.

The Tennessee Valley Authority is perhaps the first evidence of utilizing knowledge and mechanical ingenuity upon any other basis than that merely of making profits or waging war. In that great and workable plan, not only have the conveniences and profit of man himself been considered, but the experiment, already brilliantly successful, has considered first of all Nature itself and the means by which man is able to live on this earth—the forest, the topsoil, the rivers, the minerals that lie under the earth. It is all a part of one large picture.

Man cannot himself escape from Nature. Neither can he ever subdue her or attempt to exploit her endlessly without becoming himself the victim. In this country, we have, on the one hand, witnessed the spectacle of a people destroying by exploitation as rapidly as possible the very sources of their prosperity and well-being—and on the other hand, an attempt by superficial methods founded upon socialistic principles, to set things right by experimenting with the buzz saw of economic laws. In the long run one method is as futile and destructive as the other, one tearing out the very base of the whole structure, the other further weakening a structure which has already begun to totter and shake.

The fact put very simply seems to me to be that the wealth and resources of this nation are a common heritage belonging to all of us, and unless they are considered as such and cherished we are among the doomed nations of the earth. No family heritage delivered over to a quarreling family without discipline or management could for long remain intact.

Our economic balance and welfare might be compared to a three-legged stool of which one leg represents industry, one labor and one agriculture. With any of the three legs broken or missing the stool itself collapses. If any of the three is weakened, the stool becomes itself weakened and unsafe. Of the three legs, it seems to me that agriculture must forever be the key leg. At the present time, it is the weakest of the three supports.

A sick agriculture in any nation sooner or later infects the whole economic structure with its sickness. In every depression we have ever experienced the signs of catastrophe appeared first of all in our agricultural population. Yet, in all the "postwar planning" of the past two years, agriculture has been mentioned but rarely. It is all labor and industry. More than half the population of this country lives on farms or in villages, small towns and cities largely or wholly dependent upon agriculture for their prosperity. If this great potential market becomes useless through economic illness, labor and industry may attempt to go on functioning but they cannot continue to function for long. And today, even in the midst of the so-called war boom, agriculture is still sick with a sickness which will show up violently when the war is over.

The illness comes from many things—from ill use of the land, from individual farmers' lack of independence and security, from his being forced perpetually into the position of buying at retail and selling at wholesale, from his persistent exploitation by industry, by labor and by the shopkeeper. Agriculture needs better markets, better distribution, farm machinery which is not twenty-five years behind the development of the automobile and the radio. He needs, himself, to adopt newer and wiser farming methods, to cherish his soil as his only capital, to live off his land instead of buying half or more of what he eats out of stores.

He needs better cattle so that it does not require two hundred pounds of feed to produce from two cows no more milk than one good cow would give on a hundred pounds of feed. He needs to work *with*

Nature in his farming rather than to try to fight her as he has done throughout the long, sad history of American agriculture. But above all, he needs, like the workingman, the reality and not the illusion of economic security. Until a good deal more than half our population acquires that security which is based upon thrift, self-reliance and the earth itself, we shall never know security as a nation. We shall go on having fantastic depressions, distorted and exaggerated by our own follies whether of high-pressure, installment-plan selling or of fantastic, moon-eyed economic juggling. And each depression will be followed by more and more destructive taxations, as the whole economic structure of the nation grows weaker and weaker and sinks to a European and finally to an Asiatic standard of living.

For anyone to devise a plan of how to end the vicious circle is a tall order. Very likely it requires the regeneration of a whole people, a whole nation, the acquisition of wisdom and respect for the immutable laws both of economics and of Nature. A little common sense would do no harm and a little recognition of the fact that work, and not idling, is the God both of mankind and of civilization itself, and that the seesaw of higher wages and higher prices solves nothing whatever, but only moves toward that point eventually when presses cannot print worthless money as rapidly as it is spent. The root of the sickness lies deep. Only unity of the people, statesmanship, understanding leadership and wisdom can cure it. Whether we shall ever have any or all of these depends upon the American people themselves. I suspect that we shall only get them "the hard way."

THE SUGAR CAMP AND THE GREAT SUBJECT OF MORELS

THE SUGAR CAMP STANDS ON THE EDGE OF THE WOODS ON THE Beck place, a little way from Jim Pugh's cabin and his trout pond. It is a big shed with wide cracks between the boards and doors at both ends to permit the air to circulate in boiling time and drive out the smoke and steam. Inside there is a long low brick furnace with a vat on the top where the sap is boiled down into syrup. All through the summer and autumn and winter months the vat lies turned upside down, the inside coated with the thick syrup of last season to keep it from rusting. In autumn the leaves fall and drift inside to cover the earthen floor and pile against the heaps of cordwood left from last spring's boiling. When winter comes the empty shed is frequented by rabbits and raccoons and birds which take shelter there, against the force of the blizzards which sweep down the Valley from the north.

And they come there for the food that is put out for them when the acorns and weed seeds are gone and the ground is covered with snow. Twice a week someone carries up to the sugar camp a bag containing a few ears of corn, some apples and potatoes and a little of the dairy mash mixed in the big barn on the Anson place. Both birds and animals like the mash for it has in it ground corn and oats and wheat and dried green alfalfa. It is a nicely balanced food, good for man, child or beast.

The contents of the bag is scattered inside the shed on the floor and atop the wood pile where the birds can get it easily. When that is done, you go away wondering how many eyes of wild things have been watching you on your visit from beneath the old logs or out of holes in the big trees and from branches overhead. You know they are all there somewhere near, in the myriad dark holes of refuge that exist in a thick woods. You know they are all there, for if you come back the following morning, the record of their presence is in the snow outside the sugar camp.

The tiny tracks come from all directions—the footprints of the big fox squirrel beginning nowhere, at the spot where he has dropped out of a tree and scurried toward the shed; the triangular tracks of wild rabbits coming out of old hollow logs and crevices in the rocks and holes burrowed by the ground hog for his winter sleep; the tiny footprints of the muskrat dragging his tail along behind him in the snow. He has left his pond during the night and gone foraging along the creek and up the spring run until his delicate nose told him there were apples in the sugar camp. And there are, in late fall and early spring, the prints of sleepy possums with the yellow eyes that see so well at night. And the prints of the raccoon's little hands pressed into the snow. And here and there in a crazy pattern are the delicate footprints of birds—the cardinals, the chickadees, the hedge sparrows, the warblers.

They have all been there in the night or in the early morning before man is stirring or is busy in the big barns doing his chores. When you go there in the middle of the day, the place is empty, as empty as all the winter landscape. If it is a fine day, you may hear the distant whistle of a cardinal or the call of a hedge sparrow, but that is all. It is a good place to feed wild things, for no owl can attack them there and I have never seen the tracks of a fox in the snow, perhaps because he is suspicious of the shed as a place inhabited by his most implacable enemy, the poultryman.

Often I have been tempted to go there at night with an electric torch to watch the banquet, but I have never done it so far, for fear of frightening the visitors away or making them suspicious. The story is there the next day, written in the snow and on the floor of the little shed—the teeth marks of the muskrat and the squirrel in the half-eaten apples and potatoes, the ears of corn shelled and scattered. One of the visitors carries off the apples whole and I suspect the raccoon who takes his food down to the spring run to wash it before beginning his meal. At any rate, that is where his tracks lead. And the squirrel, I suspect, is the one who likes the seeds of the apples, picking them carefully out of the cores left by the muskrat. And one of them does not nibble his corn off the cob. He shells it all carefully, leaving it scattered about the bare cob. Again I suspect the raccoon.

And all over the piles of dairy mash are the tiny, delicate tracks of the feet of birds.

I should like to see them all feeding there together—the birds, the possum, the raccoons, the squirrels, the muskrat and the rabbits—but I never shall, because the first flash of light or the first footfall would destroy the peace and the security of the whole picture. I shall only be able to read the story in the tracks in the snow, in the shelled com and the tiny teeth marks on the apples and potatoes.

All the year round until late in February the sugar camp belongs to the wild things. Then one morning when the snow begins to melt and the earth to heave and the fields and pastures to stream with water, men come to the place and take it over. There is a great bustle and activity and the evaporator is turned right side up and a big fire built under it to boil out last year's sticky syrup and leave it fresh and clean for the new boiling. Outside the shed, the big iron butchering kettle is suspended over another fire and water is heated in it to wash out the sap buckets which have been piled high in a corner of the shed for nearly a year. In my grandfather's day, the sap buckets were made of wood, and a couple of weeks before an expected run of sap they had to be lined up and filled with melting snow water from the spring run to swell and become watertight again. It was a lot of work but there was something special in the rite which is missing in these days of metal buckets. I think the seasoned ancient wood, soaked year after year with fresh sap, gave a special flavor to the syrup.

Our sugar bush lies all the way along a north slope on the side of the hill where the sandstone crops out among the big beeches and oaks and sugar maples. It is the last place on the whole farm to thaw out and so the sap runs late, but is, I suspect, all the sweeter for its lateness. The crude paths cut among the young trees and underbrush to permit the passage of a wagon to collect the sap from the trees, run up and down, dangerously, across big rocks and through miniature ravines. When the ground is thawing and streaked with melting snow and water streams everywhere, the almighty tractors can't do the job. The front rears up in protest on a steep slope or the big wheels slide helplessly around. Gathering sap is a job that only horses can do, and for my money, only horses should draw the big sled with the three-hundred gallon tank. A tractor would be a desecration.

There is something beautiful and satisfactory in the sight of the two big iron-blue Percheron mares, Queen and Sylvia, seen from a distance among the bare trunks of the big trees against the snow-streaked hillside. For me it brings back memories of pleasanter times, when living was easier and people had a chance to know each other in the leisurely comfortable way of neighbors in a Currier and Ives farm picture. There are lots of reasons why I should like to be twenty years old again, but one of the reasons I am glad I am middle-aged is because it makes me old enough to remember what living was like and what farm life was like before there were automobiles and tractors and airplanes. I am not yet old enough to sit dreaming of the past but I know that life had values in those days which are gone forever, unless someday the world begins all over again. The sight of the big Percherons among the trees brings back all those values which my grandfather knew as well as any man of his time, because values were important to him. He did not simply take them for granted. He was intelligent enough to know what a good life he had and to savor it.

And the sight of the big blue mares brings back memories of my grandfather's sugar bush where we used to boil sap round the clock during a big run. It was a big shed, bigger than our own, with bunks built in it where one could sleep or snatch a nap after building up a roaring fire under the evaporator. With him, sugar making was a kind of rite celebrating the return of spring. My uncles and my father and friends from town joined him and ate chicken and sweet potatoes roasted in the red-hot coals and they drank hard cider until the sky turned gray and then frosty blue and you went to the barn to harness the big horses and make the rounds of the buckets brimming with sap at sunrise.

There is a kind of excitement which tinges the whole ceremony of sugar making, for it is the symbol of the breaking up of winter and the coming of spring when the sap rises in the trees and the first faint flush of green follows the streaks of melting snow. The cress begins to grow in the spring run and the chickadees and sparrows to call. After the death of winter it is rebirth, the beginning of hope, a new year with the promise of plenty. Even the dogs and horses feel it. The big mares stamp the earth and toss their heads and their breath steams as they snort in the frosty air of early morning or evening. And the dogs go mad, running in circles round and round the sled, chasing rabbits and squirrels that never were, save in their imaginations.

Few things on earth taste so good as the syrup of the first sugaring off. It is fresh and new, the very essence of the earth and the budding trees and the wakening spring. You can smell it in the steam from the evaporator and taste it in the hot syrup lifted from the vat in the big ladle.

Boiling down at night is the pleasantest of all. The sap boils and the steam smells of syrup and the fire under the evaporator throws shadows against the gray weather-beaten wall of the shed. On the ground, close to the evaporator, lie the dogs, grateful for the pleasant warmth, sleeping heavily after all the foolish running they have done all day. And in the corner is a jug of cider or maybe something stronger and seated on logs or cordwood you sit around and tell stories about the Valley and the people who lived there before you were born and trod this same earth that is beneath your feet and tapped the same big maple trees. A kerosene lantern casts a pleasant yellow glow over the whole scene. If you go outside into the frosty air there is the sound of the rustling and scurrying of the wild things which have come up to stare at the strangers who have invaded their territory. And sometimes, if you turn the electric torch in the direction of the big fallen chestnut a hundred yards away, the light will catch the reflection of a dozen or more eyes that gleam like green jewels in the darkness—the eyes of the raccoon and rabbits and possum and muskrat that have found the new spot a little way off, where you leave apples and potatoes and corn.

In that weather-beaten shed you are very close to the earth and to security and peace and indeed, very near to God.

This is about as good a place as any to tell about Jim Pugh and his cabin, for it stands perhaps five hundred yards from the sugar camp, visible with its little pond among the trunks of the big oaks and sugar maples. Long before we came to Malabar, Jim bought six acres of woods and rocks and spring streams from Clem Anson. The spot exists there in the middle of the farm, a little island, a kind of fragment of the past, with a cabin made like the cabins of the early settlers, consisting of one big room with a vast loft overhead where everyone sleeps. At one end, there is a big fireplace and chimney built of sandstone cut from the ledge behind the cabin in all the shades through pink to deep red.

Jim is a big fellow, who works as an engineer for the company which supplies most of Ohio with electricity. He is a veteran of the last war and loves the woods and animals and people, and Jim is an artist and a creator, for these six acres are his creation, as much a work of art as any painting or church or symphony. And Jim made it all himself. His mind conceived it and his hands executed it all.

The cabin is built of old cedar poles which have seen their best days standing along the roads to carry the wires that supplied Ohio's homes and great factories with light and power. Jim designed the cabin and built it, aided by his many friends. And beside it he built a dam to convert the waters of two springs into a good-sized pond where he, his family and his friends can fish and swim. One of the springs comes out of the ledge of sandstone above the house and the other gushes forth from the very roots of a great beech tree marked with the symbols and initials of lovers now middle-aged or old or dead, lovers who wandered there to be alone, away from the picnics held in the catalpa grove or the old mill in the Valley below.

All around the cabin grows a carpet of ferns and partridgeberry and in May when the leaves of the big oaks are pink like flower blossoms and the maple and beech leaves are lettuce green, all the space beneath is white with the blossoms of dogwood and scented with the heavy perfume of wild crab blossoms. The cabin stands a little way from the rocky lane that leads up to the wild beauty of the Ferguson place.

There is a kind of jewel-like perfection about the six acres with its cabin and pond, that kind of ordered and natural perfection of disorder which in a garden is the most difficult of all effects to achieve. Jim has created that too with the taste of a great gardener, for in all the planting, in all the perfect *wildness* of the place there is nothing to suggest that it was planned. It is a kind of small paradise where Jim and his family come for week ends in winter and to spend long happy weeks in summer.

But best of all perhaps is the fact that Jim has made the paradise, not for himself and his family alone, but for friends as well. He likes people and he shares all the delights of the place with less fortunate friends who have no escape, no refuge like this one. Jim is always having pig or steak roasts or dances. And each year he takes over the sugar camp for an all night sugaring-off party.

The cabin and the six acres of woods and spring, pond and rocks are a warm spot on the farm. Each time I pass it I feel like taking off my hat to Jim and the work of art he has created and the friendliness of the spot. I have never seen on earth a pleasanter place than Jim's cabin on a moonlit winter night with the pond frozen for skating and the ground covered with snow and twenty people sitting down to a meal of steak or pig roasted in the big fireplace and afterward dancing by the light of the big fire and kerosene lamps swung overhead.

I don't see enough of Jim and his family and the reason is the dogs. Jim has a big collie and my own dogs can never be persuaded that he is not an intruder on their private property. Since I can never leave my house without four or five of them, I have to skirt Jim's cabin to avoid a pitched battle. Only at night when they are shut up can I visit Jim's place.

Behind the cabin lies the steep jutting outcrop of sandstone where a million years ago the weight of the great glacier cracked the strata into long high caves and crevices. Above lie the thick woods and the groves and pastures of the Ferguson place where the morels grow in spring.

Like sugar making, the arrival of the morels marks a stage of spring and the hunt for them has long since become a sort of rite among all the country people of the Valley. The morels come not at the turn of the season but in May when the woods and thickets and orchards and pastures are at their most beautiful. There is no special date for their arrival since like the flow of maple sap, it depends upon the season. If you are an old morel hunter you smell the arrival of the fungi in the air. They appear as the last of the Dutchman's-breeches and trillium are fading and the May apples have not yet begun to flower.

I am always astonished at the number of people who are ignorant not only of the delicacy of the morel but even of its existence, although in parts of the East and northwestern states, the morel is fairly abundant. In France, where certain of the family grow, their season is an event in the year of the epicure and some eat them twice a day during the week or two they appear in woods and orchards and pastures. For myself, I place them at the very top of all delicacies, above *pâté de foie gras* and salmon trout and *écrevisse*; far above the ordinary well-known mushroom whose flavor is strong and coarse by comparison. Although country people continually call them mushrooms, they are not proper mushrooms but belong to another species of fungus and according to the very sketchy three-line account in the Encyclopedia Britannica they are called *Morchella esculenta* and belong to the genus *Ascomycetes*.

To any but an expert on fungi none of this means very much and I suspect from the immense lack of information on the subject even the experts know little about them. Certainly I have been unable to find any description of the five different varieties or forms which come up each spring at Malabar. Four of the forms bear a resemblance to each other and may be simply variations altering their form and color according to the amount of sun and moisture or the quality of the humus out of which they grow. All of them are gray or yellowish-gray and resemble a cluster of brains or tripe borne upright in the form of a conical cap on a squat hollow stem. One variety, the earliest, is small and firm and rubbery in texture and is found only in our bluegrass pasture where the forest had been cleared away. The next to appear is a yellowish-gray fungus of the same form but more yellow and loose in texture. They seem to flourish in old apple orchards. Latest of all come the yellow and black giants which resemble the others and are found only in the deep woods where the shade is thick and usually in the vicinity of ash trees.

There is a fifth kind which may easily be a totally different species. It grows in deep woods and thickets, a great hollow white stem capped by the prettiest of tiny cones of a rich brown striped with black. These are known locally by the homely and descriptive name of "dog pecker." All this observation is at first hand and scientific as far as it goes, and I have been unable to unearth much more information concerning them. From my observations, the "dog peckers" and at least two of the variations exist only on this continent. No one has found any means of growing them artificially and this fact increases not only their value but the excitement of the epicure during the few days when they are available.

The hunt for them during a week or two in May is one of the excitements of the year which engages everyone on the farm from the grandmothers and full-grown men to the smallest boy. There is always an acute rivalry about the finding of the first morel, and the boys sneak away from the plow into the woods and my mother goes poking about the old orchard with her stick in search of them days before they are due to appear. Then one day someone announces he has found the first ones and produces three or four in a hat or out of the pocket of a shirt. And the hunt is on!

The peculiar excitement of a morel hunt is not quite like any other excitement I know. Perhaps there is in it something of the beauty of the season itself when the woods and orchards and pastures are moist and damp and dripping and smell not only of wild flowers but of the clean tangy scent of decaying leaves. And there is the bright weather of May when the warm showers fall on you, soaking you through, producing no feeling of discomfort but on the contrary one of great satisfaction,

as if you yourself somehow drank in the warm shower through your skin and had become a part of the whole rebirth of the world. The morels are very difficult to see, for in the sun and the shadow of the decaying leaves and grass from which they spring, their range of browns and grays serves to hide them as a leopard is hidden in a jungle. Each one of them is a beautiful thing in its delicacy and moistness, beautiful to look at and smell and touch. There is something mysterious about them and the quickness of their sudden appearance. I have returned after a couple of hours to a spot where I have picked every morel to find a whole new crop sprung up. Bob says, half jokingly, half in earnest, that he believes they spring up the moment your back is turned.

For a few days, while they are in season, everyone at Malabar spends an hour or two a day looking for morels, and for these few days everyone eats morels with steak or broiled or creamed on toast. Sometimes if the season is cool and dry, they are scarce and perhaps only a handful are found during the whole season. In the springs of 1943–1944, which were warm and wet after cold, dry winters, they appeared in abundance over all the country.

The common field mushrooms grow rankly at Malabar and come in scattered drifts across the high pastures when the first fall rains begin after the heat and dust of August. They pop up their tiny white caps among the bluegrass that has begun to grow again with the coming of moisture and cool weather. The trick is to catch them out when they are young, no more than buttons or with their gills still a pale flesh pink color. They are nowhere near as delicious as the morels, but their coming too marks a season. Finding them among the bluegrass of the high pastures marks the turning of summer into autumn.

Sugar making, the arrival of the morels and of the autumn mushrooms all become seasonal rites, small feast days that are important in the year-long life of a big farm. There are other seasonal events, like the harvesting of the black walnuts, the hickory nuts and the most delicious of all, the butternut. There is the first butchering, when the air turns crisp and cold enough to keep the fresh-killed meat firm and cool, and the making of apple butter when all the farm turns out to peel apples and boil them down in cider in a great copper kettle swung over a wood fire out-of-doors. You can make apple butter on an electric stove but you won't get the same kind of apple butter. There will be something missing—the taste of copper and wood smoke and perhaps the laughter and jokes and the red cheeks that go with the ceremony. All these rites are important in the life of a farm and the individuals who live upon it, for they are a part of the very life of mankind, going back remotely into the dimness of the ages. The sense of joy and celebration is there in the hearts of all, although often enough they no longer know why it is so. They belong to what is deepest and eternal in man—those things which seldom touch the thin lives of harried city folk.

THE FRIENDS OF THE LAND

IN THE THIRD YEAR AFTER THE FARM WAS ESTABLISHED THERE appeared a long procession of cars, more than two hundred of them, advancing slowly up the valleys and among the hills of the Muskingum River watershed. At Malabar we had arranged with Hays, the caterer of the big Westinghouse plant at Mansfield for a lunch on the lawn for one hundred and fifty people. That was the first count we had received by telephone from Columbus, where the first three-day annual national meeting of a new national organization called the Friends of the Land was being held. Those attending it were coming to see what had been accomplished at Malabar, and we had invited them for lunch. Everything was ready—cold ham, shrimp salad, potato salad, ice cream, cake, coffee, sandwiches, beer and cold drinks. It was all there, ready on the big tables set on the lawn under the old black walnut tree between the house and barn.

But before the procession of cars bound north from Columbus arrived, other cars began putting in their appearance, singly, sometimes in pairs or threes. They bore license plates from all over Ohio and from neighboring states—Pennsylvania, Michigan, West Virginia, Indiana, Kentucky. Out of them stepped people who were strangers and yet not strangers at all, because all of them had the same interests as those of us at Malabar—farming and soil, water and forests. There were government people, chamber of commerce secretaries, farmers of both city and dirt variety, schoolteachers, officials of labor unions and industrial establishments.

Presently, as we began to count, panic seized us—there were already more than a hundred and fifty guests before the official procession of cars put in its appearance at the distant turn of the Pinhook Road. Then, as we counted car after car turning the Bailey corner and coming up the long road on the far side of the Valley, it became apparent that a great many more than a hundred and fifty people were in that procession. The panic grew and I went into conference with Hays. It was clear that we should have to provide lunch for at least twice as many people as we had counted on. Hays is not an easy man to daunt. He had helped take care of two thousand people who came to Malabar for a British War Relief party. He thought twice and then sent his two trucks into action. They set off at full speed for Mansfield to bring back more hams, more shrimps, more chicken, more ice cream, more forks and knives and spoons.

The cars still kept coming but Hays saved the day. In the end, five hundred and four people had lunch and saw the measures we were taking to save soil, water and forest.

The story in itself is only important because it serves to introduce the Friends of the Land and to symbolize the way the society has grown since its beginning a year before the overflow lunch took place at Malabar.

At that time there came together in Washington a group of unselfish citizens, alarmed by the waste of our natural resources. They met to organize a new society called the "Friends of the Land."

Knowing that whole regions of our country, once incredibly rich, were on the verge of becoming deserts, these men—forestry experts, industrialists, doctors, government officials, writers, bankers, professors, farmers—resolved to educate the American people to the danger. They knew our soil was being destroyed, our forests cut down without replacement, our towns and farms washed away by floods, our water supply shrunken, whole areas of once fertile agricultural land turned into desert. Approximately one-fourth of our good rich American soil had already-been destroyed, they realized, with another fourth rapidly on its way out. These were facts about which the greater part of American citizens knew nothing at all. The newly organized society proposed to inform them.

These men knew that once prosperous cities were on the verge of becoming ghost towns, not because the gold or silver had been mined out, but because the soil itself had been mined away by greedy, careless and unintelligent farming methods or because the water supply for homes and industries was rapidly failing. They knew that country banks were closing because thousands of once rich farms no longer had money to deposit or were no longer worth anything as security for loans. They knew that even in normal times we had nearly nine million "Okies" on our roads—homeless farm workers and their families on relief most of the year, living miserably the rest of the time because their farms had been mined out or eroded to a point where they could no longer provide even a miserable living.

They knew that great industrial cities like Youngstown and Cincinnati were faced with desperate water shortages and that the growth of other and smaller cities was arrested because there was no more water, not only for industry but even to supply the bathrooms and drinking water supplies of a potentially increasing population.

In May, 1942, the great steel mills of Youngstown were within two days of closing down, war or no war, because there was no more water. Exactly a year later they were within twenty-four hours of closing because of flood water. Both conditions arose principally from bad soil, water and forestry practices. In Cincinnati the water table of underground water has fallen eighty feet in twenty-five years and continues to fall at an increasing rate of speed. In Philadelphia, as in countless other American cities, the problem of drinking water has become acute. In Iowa, richest of our agricultural states, the southwest portion of the state has been transferred from rich land into the first stages of becoming a desert. A great city like Des Moines, dependent almost entirely for its wealth upon agriculture, can become a ghost city within another two or three generations unless the destruction of Iowa's topsoil is arrested. The thick, black silt which turns the near-by Raccoon River into a stream of oily muck during a heavy rain is not just earth flowing down a river; it is the banks, the department stores, the shops of Des Moines itself. In Southern states, notably Alabama and Georgia, the problem of erosion and gullies became so desperate a few years ago that landowners asked the Soil Conservation Service to come in and

take over from 60 to 75 percent of the land in a last-minute effort to save it from destruction.

As they faced these indisputable facts, the men who gathered in Washington saw that our country was rapidly approaching that point in the destruction of its natural resources which China reached more than a thousand years ago when its decline as a great nation began. They knew, in brief, what they believed every citizen should know—that with soil and forests and water supply going fast we, as a nation, were well started along the path of decline which ends in eventual destruction.

Most of the men present at the first meeting were aware, as Theodore Roosevelt and Gifford Pinchot had been long before them, that no remedial measures would be effective if *imposed by government.* They could be brought about only through education. The people themselves had first to understand the tragic gravity of the situation and then give their co-operation freely in demanding and carrying out reforms. Action in a democracy they believed must come from the people and not be imposed upon them. The states themselves must demand that measures be taken and pass the necessary laws. It was a slow process, but the only sound one under democracy. And so the Friends of the Land came into being.

Columbus, Ohio, was chosen for the first national meeting because a great many of its public-spirited citizens were sufficiently concerned about the problem to give money and time and energy to the cause of conservation; and because Ohio was unique in its almost perfect balance between agriculture and industry. There, as everywhere in the country, the problem of waning natural resources directly affected banks and insurance companies and industrial workers as well as farmers. Seventeen deforested counties in the southeastern part of the state had become public liabilities. The great industrial cities of the state were victims not only of water shortages, which limited their prosperity, but of disastrous floods which annually caused millions of dollars of damage. Indeed, the rich state of Ohio was in the first stages of a slow but inevitable process which brought about the great desert areas of countries like China, India and Mesopotamia.

The Friends of the Land had modestly hoped that they might be able to secure a respectable attendance at that first meeting. Actually, every session overflowed into corridors and galleries and received the spontaneous co-operation of the press, of women's clubs, of Farm Bureau and grange, of labor unions, of sportsmen's and business organizations.

The program included talks by authorities on water, erosion, forestry, soil conservation, economics and other subjects related to the relation of natural resources to national and local economy.

The second day of the meeting was devoted to a tour of the deforested and eroded areas in the southern part of the state to see at first hand the effects of our long policy of waste—good farm land recklessly turned into desert by strip-mining coal operations, scrubby underbrush where once great forests had flourished, wretched, inadequate schoolhouses, men, women and children so badly nourished that both intelligence and energy were permanently blighted—in short all the evidence of a countryside which had been wrecked, pillaged and destroyed.

To the visitors on that tour, something which has long been evident to experts became a reality—the knowledge that, owing to the reckless destruction of natural resources, our nation has already passed the peak of its incredible God-given wealth and is on the downgrade. Many understood for the first time that high taxes, a depressed standard of living, unemployment, poverty-stricken populations and many other terrible national problems had come into being, not through any temporary economic depression, but because the steady and reckless waste of our soil, our water and our forests was slowly bringing about a serious lowering in our national standards of living.

All this existed in Ohio, but it also existed in one form or another in every other state. In Michigan, there is a man-made wilderness of hundreds of thousands of acres. The rich southwest portion of Iowa and large parts of Missouri, Kansas, Illinois and Indiana have been nearly destroyed. In the South and West, whole states are threatened by the eventual economic and social bankruptcy which follows erosion and bad single-crop agricultural practices.

But side by side with the picture of desolation and waste those who went on the Friends' field trips were shown what could be done to check it. A field tour was made of flood-prevention and soil-conservation projects. Men and women from all walks of life visited the R.C. Bluebaugh farm in Knox County, reclaimed from eroded hills and turned from an economic liability into one of America's richest farms in the short period of seven years. The expedition visited the fourteen great dams of the Muskingum flood control area which had paid for themselves in a single year in savings to taxpayers and insurance companies. It saw the great Mohican State Forest, reclaimed from wasted land, a beautiful recreation area which within a few years will not only pay for itself but provide revenues to reduce taxpayers' levies in all parts of the state. None of these projects were in any sense a waste or reckless spending of taxpayers' money; each one of them was an investment for state and nation, showing immense profits.

On that first tour and at that first meeting scores of converts to conservation of our natural resources were made. From among the scores, several dozen at least returned home to become evangelists, some very nearly fanatics. They went back to their own communities to rouse interest in conservation programs. The Friends were besieged with requests for speakers. That first meeting gave the organization an idea of the magnitude of the job and the interest with which its efforts would be received. The Friends were on their way.

A second national meeting was held in St. Louis in 1941 and it followed a similar pattern, with one day given to business and talk and two days given over to field trips. On one of these tours the visitors saw a large area known as the Kingdom of Calloway, once one of the richest cattle-raising regions in America. During and just after the last war, when the price of corn rose to fantastic heights, this county was plowed up to raise corn. Bad, greedy practices ruined it within a period of ten years, until today many of the farms stand empty and much of the land is assessed at no more than $3.00 an acre by the tax assessors.

Later meetings were of impressive proportions, organized on a regional or state basis; others were small gatherings built about a single speaker; one of the Friends or an expert recommended by them. New members from every part of the country, all with a crusading look in the eye began to participate in the activities. They corresponded with each other and with headquarters, arranged new meetings or secured an emphasis on conservation in their schools, churches, Chambers of Commerce, Rotary Clubs and women's organizations.

The remarkable thing was that many of the converts had never before been closely associated with the land. As one man said, "I never really saw a landscape before. All my life I have been traveling around the country and what I saw from a car or a train window was just fields and roads and trees. Nowadays the landscape is alive and filled with passionate interest. I see gullies and poor farms and undernourished people and the evidences of drought and flood. Now I can read a landscape and know just how secure or how miserable are the people living on it. I worry now. I keep feeling something must be done. I'm going to help all I can."

This point of view was typical of many new recruits who gave their energy, money and time. There were many rewards, not the least being contacts with other people like themselves. Perhaps the greatest recompense came from learning to know America in a new way, literally from the ground up. People who had never been west of the Appalachians or rarely out of cities, found themselves visiting forests, farm projects and areas of devastation in Missouri or Iowa and Tennessee. The Friends are not only serving the cause of conservation, they are also making good citizenship and love of one's country a practical reality.

Probably no organization in America has carried on with so little expenditure of money. Today the headquarters are located in Columbus, Ohio, where space and secretarial help has been contributed by the Ohio State Chamber of Commerce. Active members who are able, pay their own expenses, and give their time as speakers and officers, not only to the society but to other organizations interested in one phase or another of conservation. Occasional deficits in connection with the publishing of the society's magazine, *The Land*, are met by contributions. Gradually, through a growing membership, the organization has risen above the deficit level.

Great emphasis is laid upon active participation so that all who join will be working members in their communities. Each one does his share in the general educational program which is the society's principal aim.

With the outbreak of the war, it appeared for a time that the existence of the Friends was threatened, because its most enthusiastic workers soon were called into government service or found themselves on civilian committees aiding the war effort. It was decided to reduce activities to a minimum. But almost at once a strange thing happened—the Friends of the Land was not permitted to die. Headquarters was bombarded with letters of protest as well as letters requesting help in organizing new meetings all over the country. The Secretary of Agriculture wrote the society urging it to continue and rating it not as a civilian activity but as war work of permanent, long-range value. It became apparent that the society could *not* die, even if it chose to.

The remarkable thing is that the organization continues to grow even in time of war. Perhaps the reason is that, with more and more people, the land is becoming a kind of faith of the utmost importance.

Food shortages have also brought home to many people who knew little or nothing about it before now, the relation between their food supply and the land. At the present time the organization has over five thousand active members with a growing number of local chapters.

As the society has grown and progressed, its program has in a way formed itself along definite and practical lines. Its primary purpose is still the education of the public regarding the problems of our natural resources, based upon the conviction that these problems affect every citizen regardless of his position in our economic or social system. The slogan of the organization might well be: "The civilization of this nation is founded upon about eight inches of topsoil. When that goes, civilization will go with it."

The society urges the adoption of the practices advocated by the United States Soil Conservation Service—such things as contour plowing, terrace ditches, cover crops, strip-cropping, fencing of cattle out of wooded areas, disk-plowing, reforestation, farm ponds, among many other practices. Some of these terms may need clarification for the lay man. Contour plowing means quite simply what it says—plowing along the sides of hills or rolling land so that instead of running up and down hill, plow furrows are as nearly level as possible from one end to the other.

A furrow running up and down hill becomes inevitably a gully down which water rushes causing floods and carrying off topsoil with it. A contoured furrow, level from end to end, becomes a tiny canal or lagoon holding the rain where it falls and permitting it to seep into the ground.

A terrace ditch is simply a contoured furrow on a larger scale—perhaps three feet deep and ten feet across following the slope of the hill. It is used where there may be heavy runoff water after rain. The runoff is caught in the ditch—sometimes hundreds of thousands or even millions of gallons—and held there to seep into the ground instead of rushing off into the nearest stream, carrying with it tons of topsoil.

Cover crops are crops such as wheat and rye, planted primarily to keep bare ground covered against wind or water erosion during the winter season. Such crops are fibrous-rooted and the rootlets hold and anchor down the soil and trap and hold the rainfall until the soil absorbs it.

Strip-cropping is the use on long slopes of strips of sod crops—hay, alfalfa, etc.—alternating with the strips of open cultivated land. The strips serve as checks to runoff water and soil being carried away. If runoff starts on one of the cultivated strips it runs into the strip of sod below, where the flow of the water charged with topsoil is checked and both water and topsoil are absorbed by the heavy grass cover.

Fencing cattle out of wooded areas serves two purposes—to allow forest seedlings to grow into trees and to check runoff water. Cattle pasturing a wood lot will kill off all forest growth and frequently reduce vegetation to a point where erosion begins and gullies are started.

Disk-plowing does not turn over the soil, burying all rubbish and leaving the soil bare to be blown away by the wind or washed away by the water. It breaks up the soil to the required depth, chopping rubbish or manure *into* the surface, thereby anchoring it.

Reforestation and soil forestry practices are exactly what they imply—the replanting in trees of areas which have already been largely destroyed by bad farming practices. Reforestation not only turns ruined land into an economic asset to the taxpayer but the presence of trees stops erosion, checks water runoff and floods, and serves as a valuable windbreak in areas subject to wind erosion.

All of these practices serve the two vital and primary purposes—to check the disappearance of our topsoil and to arrest the escape of precious rain water into streams, causing floods and carrying off soil. As a general rule, there is no less rainfall in the United States than there was a hundred years ago, but today much of the rain which falls simply runs off the eroded and treeless surfaces as off an asphalt pavement into the ocean. It never reaches the underground water tables so vital to our economic, industrial existence and to the life of our towns and cities.

But the story of the effects of our vanishing soil and water is endless. The Friends of the Land are fighting to spread knowledge and information on this greatest of our national problems through meetings, examples, pamphlets and books, by frank lobbying for soil, water and forestry conservation laws in Washington and the individual states and by establishing local chapters of the organization. Lately the society has undertaken to supply a glossary of information and terms regarding conservation practice as well as a list of the large number of excellent books dealing with all aspects, economic and social, of the problem. In co-operation with the Garden Clubs of America, the Friends of the Land have distributed fifty thousand copies of a "Primer on Conservation" and the demand for this pamphlet continues to increase.

The society encourages the establishment of what, in the Tennessee Valley Authority area, are known as "pilot" farms—that is, demonstration farms privately operated upon which sound conservation practices are used. These have proved strikingly effective as *examples*, more effective than pamphlets, books or even government experimental stations, for on them neighboring farmers are able to see unmistakably the gainful aspect of proper soil and forestry practices.

The society has also worked steadily for the establishment of soil conservation district acts in all the states. This is an act which enables a group of farmers to come together and set up jointly on their farms good soil, water and forestry practices. In return they receive information and practical aid from state and Federal authorities. Such acts have now been passed by all but three of the states and bills are under consideration in the legislature of these states.

The scope of the Friends' work has gradually broadened to include diet, and nutrition, stream pollution and municipal sewage disposal, decentralization of our crowded and unhealthy urban areas. Victory gardens and self-sufficiency and security through ownership of land by individual citizens—in short, to include almost everything connected with soil, water and forests.

Its aim has gradually become that of fighting for the natural and economic paradise which is this nation's logical heritage and which it could become with proper understanding and handling of its natural resources.

The officers of the society include an extraordinary variety of good citizens of every calling from every state in the Union. Its present president is Chester C. Davis, president of the Federal Reserve Bank of St. Louis. Its past presidents include Morris L. Cooke, engineer and one of the strong right arms of the Federal government in domestic and foreign economic affairs and Dr. Charles E. Holzer, one of America's leading surgeons. Among the vice-presidents are Dr. Jonathan Forman, editor of the *Ohio State Medical Journal*, authority on allergy and nutrition and Mrs. Luis J. Francke, former conservation chairman of the Garden Clubs of America. Its directors and advisers include William A. Albrecht, chairman of Department of Soils, Missouri University; James Inglis, American Blower Corporation of Detroit; J. F. Jackson of the Central Railroad of Georgia; Edward J. Meeman, editor of the Memphis *Press-Scimitar*; Paul Sears, botanist and author of *Deserts on the March*; E. J. Condon, executive assistant to the president of Sears, Roebuck and Company; Lachlan McLeary, president of the Mississippi Valley Association; Wheeler McMillen, editor of *The Farm Journal*; Karl Menninger, nationally known psychiatrist; P. Alston Waring, farmer and author, of New Hope, Pennsylvania; J. E. Noll, farmer and banker, Bethany, Missouri; Stuart Chase, economist; Hugh H. Bennett, chief of the United States Soil Conservation Department; "Ding" Darling, famous cartoonist and Nature lover; Federal Judge Robert Wilkin of Cleveland, Ohio; Isaiah Bowman, president of Johns Hopkins University and a score of others who give freely of time, advice and money to the cause of the conservation of natural resources. Former Vice-President Henry A. Wallace is an active member and speaker for the organization as well as Albert Williams, president of the Federal Reserve Bank of Philadelphia.

These are all busy people, yet all of them have found time to work in the cause of conservation. The organization has brought much richness, I think, into the lives of all of them for it has led them into all parts of America, into its forests and farm lands and villages. It has given them a greater understanding at first hand of the problems of this huge and complex nation.

Into my own life the organization has brought great richness—in the visits to foresters and farmers, in the new friends I have made; and for me the most sympathetic and stimulating people whether in China or India or Sweden or England or Massachusetts or Texas are always those who are closely tied to soil, to water, to forests. There are rich memories of a two weeks' tour of Alabama and Georgia, with daily barbecues and fish fries, and trips into the beautiful far reaches of the Tennessee Valley and among the rich cornfields of Iowa and Illinois and across the great plains of Texas and Oklahoma, where the sky is bigger than anywhere else in the world.

All of these things are the rewards of having come home, of having heeded the advice given me long ago in the green French forest of Ermenonville, of finding again the earth and the life which was always so profoundly a part of an inescapable destiny.

OF GREEN HILLS AND VALLEYS

And this our life, exempt from public haunt,
Finds tongues in trees, books in the running brooks,
Sermons in stones and good in every thing.
—Shakespeare, As You Like It

TODAY MR. JARVIS, THE BEE MAN, CAME TO LOOK OVER THE hives and put on new supers. He is the county bee inspector and we run thirty hives of bees on shares, the farm furnishing the equipment and bees, the inspector caring for them.

The thirty hives stand on the side of the hill above the Big House, sheltered from the north winds in winter and thunderstorms in summer by a ledge of pink and red sandstone rock. Time, frosts and wind have worn and pitted the sandstone and its face is covered with ferns and wild red and yellow columbine. Each winter pieces of the rock break loose and fall down the steep slope below to lodge among the wild raspberries and gooseberries that grow in the thin flickering shade of a grove of black walnut trees.

Until we placed the beehives in the grove and took to frequenting the place, a pair of red foxes had their den in a crevice that ran far back into the rock. It was a cleverly arranged home with three or four different entrances. The main hole could only be reached from above, by coming down a narrow ledge among the ferns and columbine. Twice I caught the vixen slipping delicately down the face of the rock toward the entrance, so delicately that she scarcely disturbed the foliage on the face of the rock. As she passed, brushing the ferns and columbine aside, she put down her little paws so deftly that the leaves swung back into place as she passed, leaving no evidence that there was any path there at all. Indeed, unless one stood looking down very closely there was never any evidence that an animal went up and down the face of the rock many times daily. Once I caught her returning with one of our leghorn pullets in her mouth, her tiny, shrewd head held high to keep the dangling bird from trailing across the ferns. Perhaps if I had had a gun I would have shot her, for we are, after all, in the poultry business and each summer foxes take forty or fifty pullets off the range; but I had no gun, and afterward I was glad I had none.

I have seen foxes many times but I have never seen one when I had a gun. There was a big, bold dog fox who would come across the open field in broad daylight, select a fat pullet and carry it off, ignoring your shouts and even an ill-aimed stone or two. One summer three young foxes used to sit in a row on the high ledge, silhouetted against the evening sky, their tails curled around them, not more than two hundred yards above the Big House, watching us in the garden below. I saw them there many times, but never when there was a gun handy. I do not know how they knew it but they did know. If I went back to the house for a gun, the three young foxes sitting against the sky had always vanished when I returned.

But last summer one of the dogs caught them out. Baby, one of the boxers, came down the hill carrying what appeared to be a big Maltese cat. He was inclined to be wicked with the barn cats and I thought that at last he had got old Tom who had tantalized and scratched his nose and ears since Baby was big enough to put up a fight. But when Baby came up to me, proudly, I discovered that it wasn't old Tom at all but a half-grown gray fox cub. After that, the foxes, both gray and red, left the ledge above the house. They have not been back since, save at night when in the mating season they bark and call all along the ridge above the Big House. At night they come close in so that the sound of their barking is mixed with the sounds of the sheep and cattle in the barns.

One of the most pleasant things about living in rich, half-wild country, like ours, is the feeling that when evening comes and at last darkness falls, live things stir and come to life all about us. My room, where I work and write and sleep, is a big room with a bay window and two outside doors leading directly into the garden. A little way off, the half-wild garden merges imperceptibly into underbrush and forest and the animals at night come close in to the house. In winter, on a moonlit night, I have seen as many as twenty rabbits feeding on the terrace just outside my window where we throw down grain for the turkeys, the fighting chickens and the guinea fowl. You can sit inside in the darkness and watch them nibbling at the grain, suddenly raising their long ears in sudden alarm, hopping off hysterically at the slightest sound, to return presently, in little, tentative, sudden advances until their alarm is dissipated and they begin their nibbling all over again.

In spite of the grain, they have gnawed their way around the flowering crabs and slaughtered rosebushes and young blueberries; yet I have never been able to bring myself to shoot one of them. After you have watched them like that, night after night, silently in the moonlight, something happens to you. I do not know exactly what it is save that they come somehow to be your friends, that you would feel a bully even to open the door and startle them away. Watching them, living very close to them, gives you a vague and curious sense of participating in the mystery of Nature itself, of yourself being not a specimen of dauntless, clever, all-powerful mankind, but of being only an integral and humble part of something very great and very beautiful. You feel a sudden intense and unattainable desire to step out on the terrace to speak to the timorous rabbit, to make friends with him, to talk with him there in the moonlit glittering snow. I have been brought very close on these winter nights to an understanding of the beliefs of the Jains of India who hold that the principle of life itself, even in an insect, is sacred and of God.

But for me, religion and faith have never come through churches and rarely through men. These things have welled up in me many times in contact with animals and trees and landscape, at moments when I was certain not only of the existence of God but of my own immortality as a part of some gigantic scheme of creation, of an immortality that had nothing to do with plaster saints and tawdry heavens but with something greater and more profound and richer in dignity, the beautiful dignity of the small animals of the field, of a fern growing from a damp crevice in the rock, or a tulip tree rising straight and clean a hundred feet toward the sky. It is the dignity and beauty which man managed to translate into the stone of Chartres and St. Cernin, but somehow missed in the ecclesiastical manifestations of his spirit.

In the daytime, mourning doves come to the same terrace to feed with the guinea fowl and turkeys, small silver-gray doves with rings of darker pink beige for collars. They stay with us all through the winter,

living on the dry sheltered ledges of the sandstone rock above the house. With them feed the chickadees, the song sparrows, three or four varieties of woodpeckers, the cardinals and the noisy, vulgar sparrows. They all feed together in peace, save when that beautiful and arrogant fellow, the blue jay, drops down and bullies them all away.

In the mornings, the borders of the pond are marked with footprints of the raccoon who has come down in the night to wash his face and his food before eating. A little while ago, I found among the wild iris growing at the edge of the big pond, the body of the grandfather of all raccoons. He was very nearly as big as a dog and very gray even to the spectacles which outlined his eyes. He was lying on his side, quietly, dead. He had come down for the last time to his beloved pond to die in peace of old age.

Last autumn I was awakened in the early morning, for the first time in my life, not by a sound but by a smell. As I opened my eyes, I knew what it was—pure essence of skunk. Beside my bed were three of the boxers, wriggling and shaking themselves, their eyes smarting. They had gone out of my room as usual in the early morning and on the doorstep they had encountered a skunk. The encounter had been brief, and following Prince, who is clever enough to open doors, they had returned to tell me of the encounter and ask me to do something about it. I went to the door and there on the opposite side of the ravine, I saw Mr. Skunk. He was making his way back to the shelter of the forest and he was in no hurry at all. Had he not just put to rout three big boxers? He even stopped now and then to tear at a stump or turn over a stone, delicately, in search of a fat grub or two. The skunk is an animal of great dignity because he can afford to be.

And opossums too come close to the house. I have seen them on the driveway late on autumn nights in the lights of the car apparently enjoying the warmth that has remained in the gravel. The possum is a slow-moving, lazy fellow whose glands will not permit him really to run but if he is caught on the ground he will put up a ferocious battle with teeth and claws. Not long ago, Pete, the dairyman, ran one down on foot by the lights of the jeep. He brought it into the house by the tail to show the children. The possum seemed to take the whole adventure lethargically. He was a young fellow and even when you put him on the floor he made no attempt to run away, but only nosed about the floor without any sign of alarm. Possums are comical beasts that look and act a little like clowns, and this one was no exception. Now and then he would stop moving and look up, blinking in the light, with a wicked twinkle in his yellow eyes. When you held him by the tail he would turn and lazily try to climb up his own tail and bite you. A mother possum on a branch with a whole family clinging to her back is one of the most comical sights in the world. I have seen it only once. The possum family are the only marsupials outside the continent of Australia and they have an air of belonging to another world, survivors, as they really are, of a prehistoric time.

Twice I have seen one of them actually "play possum." Once, in "the jungle," the dogs, running ahead of me, collected about some object on the ground. I discovered that it was a possum, on his back with his feet in the air completely dead. The dogs sniffed the funny yellow body. When it did not move or run, they lost interest and went away. I touched the possum with my toes. He was still soft and flexible and I thought, "He can't have been dead for long" and went on my way. Ten minutes later returning by the same path the possum was not there at all.

And another time the dogs caught a possum in the open in broad daylight. Gina was shaking him when I came up and called them off. He was a big fellow and must have weighed about fifteen pounds or more. I picked him up by the tail to carry him back to show the boys who were shelling corn by the lower barn. I was nearly a-mile from the barn but during the whole of the long walk, while he grew heavier and heavier, he showed no sign of life. His size created interest among the boys and I tossed him into the back of the wagon to be dumped into the field for the buzzards. He lay there in full sight of the boys who went on shelling corn; but all the time the possum must have kept one eye open, watching them, for when they had finished and looked in the wagon, there was no possum there. Looking about for him, they discovered him making his way lethargically across the bluegrass two hundred yards away. They let him go to return to his family and no doubt describe in possum language his remarkable adventure.

On a moonlight night you can sometimes look out of the window and see a muskrat or two swimming across the pond—a tiny dot which is his nose with a great V wake spreading out about him in the still water. If you go outside for a closer view the dot will disappear at once beneath the surface. They are shrewd and bold little beasts, very nearly as shrewd and bold as the fox. In winter they will move up from the marshes in the Clear Fork Valley and dig themselves homes in the banks of the little brook that flows only fifty feet from my window. You know that all through the night they are there, quite near you, feeding on the roots of your best iris, even coming up to the terrace outside the door to eat the grain scattered for the guinea fowl. But you never see them, or at least you see no more than a fleeting shadow, so swift you cannot identify it. Only once has one of the dogs ever killed one of them. The honor went to Lady, Max's pointer, who has a nose fine as a needle. All through the marshes in our county there are millions of them. Their fur in winter is of high quality and more than once, dyed and plucked, it has been mistaken in a theater or a restaurant in New York for sable. They are one of the reasons why Ohio ranks third among the fur states of the Union.

Cheeriest of all the night chorus is the music of the frogs. They make a kind of part-singing, ranging from the shrill call of the spring peeper to the "jug-o-rum" of the bullfrog. The tree frog, especially on hot nights, sings a kind of obbligato and the leopard frog sings baritone. Last year, a friend in the Game Control gave me five hundred giant tadpoles of the big Louisiana frog, prized for its legs. I distributed them among the ponds, dubious of their survival in the cold northern winter, but some of them at least managed to live, for this summer there is a new voice in the part singing—a deep basso like that of Wotan.

Night in our county is far from silent as many city-dwelling friends have discovered when they come to stay. There is always the chorus of the frogs and at certain seasons there is the barking of the mating foxes back and forth from the wooded ridges on both sides of the Valley, and in autumn the baying of the fox and coon hounds running the woods, a distant beautiful sound which sometimes comes near enough to rouse all the dogs on the place to a frenzied barking. And there is the discordant, squeaking-gate noise of the guinea fowl disturbed in the night or the gobble of the wild turkeys, or the hysterical cackling of the fighting chickens. They all choose to live in a tree just outside the window of the large guest room. Or from the lower farm may come the calling of the geese, disturbed by a fox or a weasel. A lonely farm never need be unguarded while there are geese and guinea fowl. I doubt that any intruder could come within five hundred yards of the house or barn without rousing an unwelcome din that would echo up and down the Valley.

But we have come a long way from Mr. Jarvis, the county bee inspector. He is a frail little man who took up bees because his health was too bad to work any longer in the shop where he made his living. That was a long time ago and the years have made him expert. He has bees of his own, placed in orchards all over the county. Indeed, bees are his whole life. To be a good beekeeper, that is necessary; for bees are a complicated business, or rather a profession in which there is a need for art. The true beekeeper has to be gentle and good-tempered. There are some people who naturally rouse the bad temper of bees and they are very temperamental creatures. They suffer from the heat, and become like humans, uneasy and short-tempered in the hot, muggy stillness which precedes a thunderstorm.

We had been having that kind of weather for days on end, the kind of weather which makes the corn grow so fast that if you stand quite still in a big cornfield on a warm night you can hear the faint crackling sound as it pushes upward and the stalk increases its circumference cell by cell. Corn has to grow fast in our country, since it is planted in late May and has to ripen by September. It has to grow eight to ten feet tall, blossom, bear ears and ripen them in a little more than three months. There is a rich, dark green, tropical beauty about corn that none of our other temperate zone crops possess and on a hot night there is a tropical smell about a lush field of corn; the air is filled with the scent of pollen and fertility.

But the weather which corn likes is not the weather for bees. They like clear, sunny weather with cool breezes and blossoms that open to the sun.

Three times Mr. Jarvis had come to our shaded hillside beneath the fox lair that week and each time sudden thunderstorms had made the bees angry and hard to work with. And now, on the fourth visit, the air was hot and still and there were thunderclouds like great heaps of lemon sherbet in the west. It was still not a good time to work with the bees, but Mr. Jarvis was worried about some of the hives swarming while he was away. He knew that in each hive there were queen cells. If the ruling queen neglected to open one of the cells, tear out the grub and murder it as was her habit, a new queen might hatch and take part or all of the swarm with her. It had happened before. We had lost swarms that went off to establish themselves in hollow trees in the deep woods. They had a liking for hollow basswoods perhaps because just outside their door in the spring of the year hung the most delectable of nectar-filled blossoms. I know where there are two of them, one above the big cave, the other near the raccoon tree in the old orchard.

So Mr. Jarvis put on his bee bonnet, rolled up his sleeves, picked up his smoke-bellows and went to work. Usually he scorns both bonnet and bellows but on this day the bees were, as he put it, "Very sassy."

On the doorstep of each hive the worker bees were standing with their backs to the opening of the hive, their feet anchored, their wings fanning briskly to force fresh air inside. Bees are tidy creatures and hate stuffiness as much as do sensible human beings. And inside the hive a great deal of work was going on. Cells were being built, some to harbor eggs to keep the population going on; honey and pollen were being stored for the winter; the eternal house cleaning was in progress. Heat or no heat, they were getting on with their work. It was no time to disturb them.

But Mr. Jarvis went to work with his bee bonnet tied about his throat and his sleeves rolled up. A whiff of smoke from his bellows, meant to stupefy and calm the bees, had very little effect. The hive was what beekeepers call "a strong one," with a big and healthy population. He lifted off the top "super" where the first honey of the season, drawn from the spring wild flowers, the apple and pear blossoms and the blossoms of the black locust trees, was stored. Already in June it was so heavy that it taxed Mr. Jarvis' frail strength. Inside there were hundreds, perhaps thousands of bees busy at their marvelously organized work, moving about capping cells, feeding the young grubs that were to become drones and workers. Somewhere among them was the queen surrounded by the cabinet of workers who constantly attend her.

Prince and I sat at a little distance watching. Prince, who would not leave me even when I went among the thirty hives, was uneasy with eyes and ears cocked against the assault of any angry bee. Despite the heat and the distant thunder it was pleasant there on the hillside among the ferns and columbines and wild raspberries. And there was pleasure in the sight of the hives heavy already with honey—the kind of a pleasure which in a countryman raises up pictures of long winter nights with a cellar or storeroom well stocked, of wood fires and fat cattle standing in clean straw to their knees in the great barns.

Mr. Jarvis got stung once and then again and again. They couldn't get at his face because of the bee bonnet but they attacked fiercely his bare hands and arms. "Darn!" said Mr. Jarvis, "They're really ornery today."

Beekeepers say that after you have been stung many times you don't feel the pain in the same way greenhorns feel it. It must be true, for Mr. Jarvis, as he worked, taking out comb after comb to look for queen cells to destroy, was taking a terrible beating without any evidence of discomfort.

Then, as we watched, a dive-bomber bee came out of nowhere and landed on one side of my head. There was a sharp pain followed by a burning, itchy sensation. I had been stung before when I helped Mr. Jarvis. It wasn't exactly a pleasant sensation but I wanted to learn about bees and anyone who has ever had to do with bees knows that one doesn't learn overnight. Nor can you learn out of books. Any book on bees requires as much study and concentration as a whole college course and even when you have finished, you are nowhere. A great many people believe that to have unlimited quantities of honey all one has to do is set up a hive and place a swarm of bees in it. It is not like that. I have the same respect for Mr. Jarvis that I have for a great scientist. He knows about bees and that is something—something which can't be learned out of books.

Then as I watched, another dive-bomber struck Prince. He yelped and rolled on the ground, snapping and biting at his nearly invisible attacker; but he did not go away. Then one struck me on the nose and another on the throat. One entangled himself in my hair.

Mr. Jarvis said, "You'd better put on a bonnet. I'll get you one from the car. I'm going to give them another good whiff of tobacco smoke. Fun is fun but I've had enough."

He must have been stung twenty or thirty times. With each clap of thunder the bees seemed to grow angrier.

I thought that with a bee bonnet I could watch more closely but as soon as I put on the bonnet a strange thing happened. My eyes began suddenly to stream and my face to itch intolerably. I could feel my features losing their shape. I was ashamed in front of Mr. Jarvis, who went calmly on getting stung, so I said nothing at all. But the sensation became more and more unbearable and I could feel it spreading downward from my face and head into my shoulders and arms. I wanted suddenly to throw myself on the ground like Prince and roll among the ferns. At last it could be borne no longer and I said to Mr. Jarvis as casually as possible, "Well, I must get back to work." I took off the bee bonnet and turned toward the house.

The hives are not more than five hundred yards from the house itself but by the time I reached my room, the itchy burning sensation covered the whole of my body. I could feel it spreading downwards from my head as the blood carried the poison through me. It was an extraordinary sensation, like some subtle torture, as if you could feel every artery and vein and nerve throughout the body.

When I looked in the mirror, my eyes were bloodshot and watery and nearly swollen shut, my nose swollen to twice its size. When I undressed, my body was covered with red welts. Quickly I had a bath with the green soap which was kept in the house for poison ivy. It did no good and then suddenly the shock of the poison brought on a violent chill. When that had passed, there came into my mind from somewhere a forgotten piece of knowledge—that bee venom was highly acid and that probably it was too much for a system which naturally suffered from acidity, and so I took a giant's dose of bicarbonate of soda dissolved in water.

Just as I had drunk this, Bill Windsor, the fish and game manager of the Conservancy appeared. Clad only in a silk dressing gown, with my eyes swollen nearly shut, I met him. He took one look at me and said, "Bee stings, eh?" I told him about the red welts that covered my body. "Hives!" he said. "Some people can't even eat honey without getting hives." I told him about the bicarbonate. "That was right," he said. "Go and take some more. It neutralizes the acid."

In an hour the itchy and burning torture was gone completely. My face, however, remained swollen and the next day, one eye had a beautiful shiner.

These thirty hives of bees play a large part in the economy of Malabar. They cost us nothing beyond the original investment, for Mr. Jarvis tends them and we share the honey. When for a few days, sugar became unobtainable, it did not matter to the families of Malabar, for honey and maple sugar took the place in baking cakes, in tea or coffee, on breakfast food, in all the countless uses for sugar. When supplies of sugar are normal, it is more profitable for the farm to sell its honey and maple syrup and buy sugar.

But the economic benefits do not end there. The thousands of bees work for us in pollinating fruit—the apple and peach and pear and plum trees, the strawberries and the raspberry blossoms for which they appear to have a passion. From plant to tree, from blossom to blossom, they go dusting flower after flower with the pollen carried on their tiny furry legs. They work the wild white clover, the Dutch clover, the Ladino, the alfalfa, the alsike; pollinating flowers which otherwise might go sterile, and slowly as the years passed, the evidence of their work showed up not only in yields of clover seed and the thickening of pastures and meadow growth, but along the roadside and in long-dead gullies where the seed of white, Dutch and Ladino clover drifted mysteriously in and started legumes growing where before there had been only weeds or bare ground. They played a big role in the balance of Nature which we were endeavoring to set up again on poor, wrecked land.

Their big cousins, the bumblebees, went to work for us too in ever-increasing numbers on the mammoth and the red clover and upon certain small fruits. Ecologists have long since established the fact that bumblebees, like pheasants, shun poor, worn-out land, and their population increases as land is returned to fertility. Taking a census of bumblebees is not a simple and easy operation and we have never attempted it, but in the plot consisting of the flower garden and the adjoining raspberry plantations their numbers have unquestionably increased as much as tenfold. And when the fields of red and mammoth clover are in blossom, the bumblebees work there in platoons and whole armies. On a still day you can hear the drone of their buzzing chorus a hundred yards away. They, together with proper land use and steadily growing fertility, have increased the yields of clover seed as much as 200 per cent.

But there is nothing remarkable or startling about all this. It is a part of the whole balance of Nature, from the bacteria and earthworms working deep in the topsoil to the crests of the ash trees high above, which no longer die out because there is plenty of moisture and underground water.

Slowly, week by week and month by month, year by year, the whole of the landscape about us has changed, imperceptibly at first, until now, as this book begins to draw near to the end, it seems almost a new world. The same outlines are there, the same soft contours, yet there is a difference. The thin, half-starved look is gone save for a few bare hilltops on the Bailey place.

What was once the hog lot and swamp just below the Big House is a garden now with a clear spring stream running through it all the year round—the spring stream which once flooded and tore out roads and bridges after a heavy rain and dried up during the hot months of August and September. Where once there was a gullied hillside, there grow now a multitude of flowering shrubs and in spring the whole hillside is bespangled with the white and yellow of narcissus and daffodil.

Even during a cloudburst, the water no longer rushes down across that slope from the old orchard and steep pasture and cultivated fields on the little plateau above.

The story of what happened there is, on a small scale, the story of all four of the farms. It begins really at the very top of the hill on the little plateau above the Big House by the orchard. Once, after a rainfall, the water poured down the rows between the corn across the steep little pasture, cutting its way through the old orchard below and finally across the steep slope that rose above the swamp and hog lot. These floods never occur any more because on the little plateau the row crops are contoured and the water stays and sinks into the ground. After this year, there will be no row crops at all on the little plateau above the Big House but only hay and wheat grown in rotation—hay year after year until it thins out and then wheat in order to seed it back again into hay. On the steep slope of the little pasture the scars of the old gullies have seeded themselves over with bluegrass and white clover. In the old orchard where once crops were grown and the soil left bare all winter to wash away with the winter rains and frosts and thaws, there is a thick sod of orchard grass and between the rows of hundred-year-old apple trees, grow peaches, pears and plums and grapes—Concord and Niagara, Delaware and Moore and golden muscat. And in the half shade grow also red raspberries, healthy and extravagant, and blueberries and blackberries, all mulched but never cultivated. No water cuts across it now, no topsoil slips away imperceptibly yet by the ton if one took the trouble to measure it.

On the steep slope above the garden, the ugly deep gullies are healed across with growing shrubs and flowers, day lilies and peonies and poppies and iris. The steep hillside took care of itself once the runoff water was stopped and held on the land above. That slope has never been cultivated. Nothing was ever done to the flowers and shrubs and trees which grow there save to mulch them heavily with barnyard manure once a year in November. Each year the mulch was left on the ground and gradually beneath it, in the earth which once in summer dried out into a hard bank of yellow clay, there grew up whole colonies of earthworms, hundreds of thousands of them which fed on the decaying mulch and the grain in the manure. Year by year it was possible to see the clay bank change into topsoil, imperceptibly yet with extraordinary rapidity. On the surface there grew up first of all a layer of dark loam soil and, beneath it, the clay itself, permeated by the slow infiltration of manure water and churned over and over by the action of the worms, began to disintegrate and grow soft and loose and a little darker in color. It was rich enough, that glacial clay; it needed only mulch and humus, bacteria and earthworms to turn it into the richest of soils. At any time of the year you can dig beneath the mulch and sod and find the soil loose and moist. And all the moisture does not come from the rain which falls directly upon it. It comes, much of it, from the water caught and held on the little plateau a quarter of a mile away and nearly two hundred feet higher up, and from the rain held by the sod and mulch among the grapes and apple trees and raspberries and small fruit trees of the old orchard just above. On what was once a bare clay bank grow today the finest flowers in the garden, healthier and finer than the flowers in the cultivated, carefully prepared borders in the flat part of the garden below. On two spots of the once dry bank there are seepage springs from which water oozes all summer long.

The bluegrass and white clover of the steep little pasture below the outcrop of pink sandstone grows lush all through the dry months from the water which instead of rushing across its surface, seeps through the ground from the little plateau above.

In the flat garden, the little stream no longer goes wild and tears out small trees and great clods of sod. It flows as it must have flowed when the first settler came into the Valley, limpid and clear and steadfast because the springs above which feed it are in turn fed by the trapped rainfall which no longer uselessly flows off the bare land to flood the Valley below.

The new pond, below the old still pond which silted up with the good soil from the hill fields above until it became, first a swamp and then a garden, is almost free from silt. It will be there a hundred, two hundred years from now, still free of any silt save that which creeps in from the near-by roads during a heavy rain. The spring over which the first settler, John Ferguson, built his little cabin now flows clear and cold all year out of the roots of the big black walnut tree.

Ours is naturally a county of springs; rolling country, with glacial moraine piled on top of the thick layers of pink sandstone. What happened to John Ferguson's spring happened to all the other twenty-odd springs on the place. As the forest was turned back to Nature, the hills contoured and stripped, the earth kept covered in winter by an even green blanket of wheat or rye, the water no longer ran off the four farms carrying with it the precious topsoil. Instead, it sank into the ground to come out again in clear flowing springs. As year passed into year the flow of the old springs increased and new ones appeared.

One morning in the third year Pete came into my room and said, "Come with me, up to the Ferguson place, I want to show you something."

He wouldn't tell me what it was he wanted to show me, but by the grin on his face I knew that it was something good. We couldn't go farther than the gate in a car because it was early April and even the old Ford had been known to become mired in the lower spots of the big hilltop pasture. We walked up over the crest of the hill where you have a view of three counties, and halfway down the far pasture in a grove of walnut trees we came to what had excited him.

There, at our feet bubbling from the very roots of the trees, was the most beautiful of sights—a new spring with a three-inch stream of clear, cold water bubbling up in the midst of the pool which had formed in the depression among the trees. From somewhere deep underground, a vein of water had suddenly forced itself up through the soil of the pasture. Once, long ago, there must have been a spring on the same spot, for near by grew the red daylilies and the pale green star-of-Bethlehem which always mark the site of a settler's cabin in our country, but as the land above had been farmed more and more badly, the spring had diminished and finally died. Then we had come to the place and anchored the soil above and covered the bare earth in winter with cover crops and turned much of the land back to sod, so that the falling rain no longer ran off the high hill above but sank into the earth to accumulate deep in the strata of the underlying sandstone. And so, as the underground reservoir grew, the accumulated water at last reached a level where, on this April in 1942, the old spring was reborn, gushing out from under the walnut trees as it had done when the first settler came upon it.

That spring has never failed since, even during the bad drought of 1944 when farmers all about us in Ohio were hauling water for their livestock. During that drought, the worst our part of Ohio had known for fifty years, only two or three small springs out of the twenty-odd at Malabar showed signs of failing. Never for one day were our cattle without water in any pasture of the thousand acres.

What we had done was a simple thing; simply to restore the balance of Nature, to keep the water where it belonged, on our land rather than turning it loose down the long course of the rivers finally to reach the Gulf of Mexico. And now in drought time we had the water we had stored up underground during the seasons of good rainfall.

There is something beautiful and exciting about a deep, clear flowing spring, even in the rich, well-watered Ohio county. It is a sight, I think, which strikes deep into the ancestral memory of man, going back thousands of years into Mesopotamia and Egypt and India to the very roots of man's beginning and civilization.

There is on the Bailey place, a famous spring, one of the largest in all Ohio, where a whole brook gushes out of the sandstone outcrop behind the old house. It flows through an ancient springhouse with great troughs hewn from single blocks of the native stone. In the troughs filled with icy water stand cans of milk, crocks of cream and butter. In late summer, cantaloupe and watermelon float there, chilled by the living water. The stone walls are damp and moisture hangs in drops from the ceiling and outside, where the icy brook flows swiftly down toward the barn to water the cattle, the steep course is choked with crisp, spicy watercress. In summer, the small boys go there to fill the big jugs they carry to men working in the fields at harvest time.

What happened in the little garden and the springs and the steep slopes above has happened over nearly all of the four farms. Now when one stands on the little porch in front of the Big House, the whole of the landscape has a look of lushness that had not been there for fifty years or more, perhaps not since those first years after the settlers cut away the great oaks and beeches and maples and began to farm the thick black virgin soil.

In the bottom pasture the bluegrass and white clover grow like an extravagant lawn, the kind of lawn an English gardener dreams of. The lime and the phosphate have made it dark green and thick and juicy and the moisture from the blocked drainage tiles and the mulch which has been built up by pasture mowing keeps the bluegrass cool and growing throughout all but a bitter drought.

Along the road at the edge of the pasture and around the barnyard of the big barn, is a row of big locust trees planted there to feed the bees. They serve not only the bees but they add immeasurably to the beauty of the landscape and in the still evenings when they are in blossom their heavy perfume drifts all the way past the bass pond up to the Big House. Beneath them the bluegrass grows sweet and lush because the locust is itself a legume and pours nitrogen into the soil. They represent neither much expense nor much trouble—a few hours' work with a spade and a few cuttings thrust into the ground.

And below, around the little bass pond on the Fleming place, where the geese and ducks have their world and are joined spring and fall by their cousins the wild ducks, the once bare banks are covered now by the feathery green of the Babylonica willow and the red-stemmed water dogwood and a few young sycamore trees. A little above the pond, there grows on land that otherwise would have been wasted, a young orchard of pears and apples. The geese and ducks nest among the forsythia that covers the banks, its golden yellow reflected in the pond.

All of this transformation involved little expense or trouble. The willows and the water dogwood came from bundles of cuttings carried under my arm during Sunday afternoon walks and thrust into the damp soil along the shallow edges of the pond where the big bass and the bluegills and the red-winged blackbirds make their nests. The forsythia came from young plants sprung up beside their parents and carried along on Sunday walks and put in with a child's spade. The sycamores seeded themselves and were not cut down ruthlessly by a man with a scythe and more energy than intelligence.

That little corner with its bass pond had been bare as the head of a bald man when we came up there. The same tenants who left piles of old tin cans and rubbish outside of their door, took the trouble to scythe it clean of all growing things once each summer, for what reason neither God nor man can, I think, divine, unless it was the atavistic instinct inherited from pioneers with a fear of the encroaching forest or the precautions of those insatiable farmers who admire bare wire fences and monotonous fields.

Today the pond and the little corner in which it is set have become a place which produces fish and fruit and is pretty to look at not only for ourselves but for the passers-by on the Hastings Road. Frogs and turtles and muskrats live in it. Killdeer and red-wing blackbirds and song sparrows nest along its edges and big gray Toulouse geese and Peking and Muscovy ducks live there and breed and raise their young, feeding themselves most of the year with only a little grain thrown them in the bare winter months. It has become a little world of its own, full of life where exciting things happen—tragedies like the advent of a great and ancient snapping turtle which killed the baby ducks until one day the dogs caught him offside in the pasture and I put an end to him, and the endless comedy of the big bold gray geese who fear not to grip an old sow by the ear to drive her away from the corn, or to chase the big boxers if they come too near a nest. And the antics of the big male bass guarding a nest who will attack your finger if you thrust it into the water near him. And the breeding frenzy of the toads in the early spring in the shallow water, a spectacle which surpasses in lechery and urgency the most terrible Babylonian orgy. And there is the compensating and satisfactory beauty of the wild iris and the bulrushes, and the tragedy of the big water snakes that prey on the frogs and small fish and have to be killed by Baby, the biggest of the boxers, who inherits his propensity for snake killing from his noble and deceased father. And there is the scent of crushed mint underfoot as one walks along the edge of the pond on a hot day.

It was not much trouble, that pond and the little world surrounding it. I think I love it best of all the spots on the whole farm, better even than the lofty remote and beautiful world of the Ferguson place. Up there against the sky looking across thirty miles of Valley and stream and lakes and woods and hills, one comes close to God with that sense of remoteness and grace and insignificance which the Hindu knows, but down below in the valley by the pond, one comes very close to all that lives, to the geese and fish and birds and frogs and turtles. Beside the little pond one is no longer insignificant, a mere atom in a vast universe beyond comprehension which produces in the spirit a remoteness and peace not untouched by terror; in that smaller world one understands everything, with all its tragedies and comedies. One sees that even among geese and ducks there are braggarts and bullies, heroes and knaves, clowns and heroes. It gives one a nice reassuring feeling of belonging to a whole scheme of things, of being a part of the whole plan of Nature. One begins to understand exactly where man fits in.

And beyond the pond and the white and green barn of the Fleming place, lie those fields which were abandoned and grown up in weeds when I first saw them again after thirty years. Now they lay in long strips of green following the soft contours of the low hills left long ago by the icy glacial rivers, green with alfalfa, with the lighter green of corn and the sea green and later the gold of wheat and oats. And, in the near end where once there was only desolation, grow the rows of potatoes, the sweet corn, the peas, the lima and string beans, which provide six families with all they need and more for all the summer and the long winter that follows.

What one sees there is a kind of miracle made by the hand of man with work and intelligence and an investment of money which has been repaid many times over. The old square fields are gone and in their place are the strips of green following the contours of the earth itself. Where the slopes are steep, grow only green hay or alfalfa or wheat which binds the soil and keeps it from wasting uselessly away. And each year that soil, anchored by trillions of tiny small roots and sheltered by the grass above, grows a little deeper, a little richer with nitrogen and

bacteria and worms and the manure that is spread over it by the wandering cattle in the autumn and by the man-made machine when the ground is frozen and covered with snow. And the fence rows along the roads and between the fields are not merely bare wire fences; they have been allowed to grow into hedgerows which check the moisture and provide shelter for birds and wild game of every kind.

Beyond the fields the distant high bare hill of the Bailey place, the last of the farms to be acquired and the poorest, have already begun to turn a deeper green as the lime and phosphorus seep into the starved hillside and disintegrate and feed the new growth of bluegrass and white clover. And each year, the cuttings from the pasture mowings will build a little thicker the blanket of mulch and humus which keeps the earth beneath cool and moist and the bluegrass growing through the long hot days where once the vicious heat on the bare, overgrazed earth burnt the roots and all but killed it.

But perhaps the woods have changed more than any part of the landscape. Only a few years ago the ground was bare beneath the high trees. There were no seedlings and no fine roots and the leaves, as they fell, blew away to pile up against the fences in drifts. The little grass that grew there was shaded and sickly. It contained little nourishment for the cattle and sheep who were expected to pick up a meager living on the pasture they found off the grass and the young seedling trees. Gullies scarred the bare hillsides where the natural process of forests was checked.

Once the cattle and sheep were kept out of those woods a transformation began to take place on the floor of the forest. The first year a whole crop of seedlings came in—tiny oaks and beech and ash, hickory and maple. And as year passed into year, the seedlings grew and new ones appeared until presently the woods began to take on again the appearance of a real Ohio forest. The seedlings grew and produced leaves which dropped to the ground and their trunks prevented the leaves which fell from their tiny branches and from the big trees which spread overhead from blowing away. They stayed where they fell, on the floor of the forest, and began again to build up layers of decaying vegetation. And when the rain fell, it no longer fell on bare earth but seeped into the ground to feed the big trees and the new seedlings. And the water trapped on the fields above the hillsides by the sod and the contours began to form seepage springs again all through the floor of the forest. The ash trees which like cold wet feet stopped dying at the tops when they reached the height of fifty or sixty feet and each year the yield of sap from the sugar maples increased. And nowadays, when one enters the woods, one enters a green jungle with a tangle of seedlings twice as high as a man's head in which it is possible to lose oneself completely during the hot, summer days when the trees are in full leaf.

One day the seedlings will be great oaks and beeches and maples and ash, tall and straight and round as forest trees should be. The time is not too far away when, harvested like crops in the field, they will bring a good return in money. Meanwhile they are serving to build new soil; they are piling up layers of decaying leaves to absorb rain as a sponge absorbs water, to check the floods which periodically sweep away good land and houses and people all the way from Pleasant Valley to the mouth of the Mississippi. And the water they absorb into the earth comes out again in the pasture and fields of the valley below to feed the bluegrass and the crops when the hot dry season comes. All these forest seedlings are a good investment, one of the best investments we have made, not only for ourselves and for the immediate future but for our neighbors and for the people downstream on Switzer's Creek and the Clear Fork, the Muskingum, the Ohio and the Mississippi. They are an investment not only for ourselves but for the nation.

One of the great changes in the landscape in our part of the Valley is in the quality of its greenness. Each year it has grown a little deeper, a little darker, a little richer. It is a new and healthy greenness born of many things, of work, of humus, of elements restored to the soil, of intelligence, of love, of water, of working with Nature instead of fighting her. The job is by no means finished. There are still places where the topsoil is too thin, where there is still not enough humus to hold the moisture, where the pasture is meager or the crops too pale a green. Many things have plagued us; many things have stood in our way—the clamor for greedy production of food, the war and the loss of man power, the difficulty of obtaining fertilizer and fencing and proper machinery, and sometimes the weather. Our progress has been slowed but not stopped. We who are interested in the adventure could not stop if we chose to for it is in the blood of all of us to make that countryside each year a little more lush, a little deeper green in color, a little more productive.

We have set about to turn the wheel of fertility moving forward again. It had been moving backward almost since the day the trees of the virgin forest were cleared off it. What we have been doing has been a relatively simple thing. We have sought merely to build as Nature builds, to plant and sow and reap as Nature meant us to do; we have sought to rebuild the earth as Nature built it in the beginning. With man's ingenuity we have been able to do it more rapidly than Nature herself, but only because we worked with the law and within the idiom of Nature. Man is never able to impose his own law upon Nature nor to alter her laws, but he can, by working with her, accomplish much, whether it is in dynamos or the airplane or the earth or the body of man himself. The man who loves Nature comes nearer to an understanding of God. Even man's religion grew out of Nature itself, and the good earth and true faith have never been removed from one another. They are as near today as they were ten thousand years ago.

The adventure at Malabar is by no means finished but I doubt that the history of any piece of land is ever finished or that any adventure in Nature ever comes to an end. The land came to us out of eternity and when the youngest of us associated with it dies, it will still be there. The best we can hope to do is to leave the mark of our fleeting existence upon it, to die knowing that we have changed a small corner of this earth for the better by wisdom, knowledge and hard work; that we shall leave behind upon it the mark left by Johnny Appleseed and my grandfather and Walter Oakes and the other good farmers or lovers of Nature of whom I have written in this book.

Not even one small part of the big program is complete for there will always be fields and fence rows and woods which can be still greener and richer and more productive of life and food and beauty. And even those fields which have been rescued from barrenness and death require eternal vigilance and the awareness that to keep them fertile and green, we must pay back our debt to them in kind. Each mistake, even the smallest one, each attempt to cheat or short-cut Nature, must be paid for in the end by ourselves.

The whole adventure has not been without disappointment and heartbreak. There have been droughts that broke the heart as the corn withered and the pastures turned brown, and there have been with us from time to time a man like Lester who broke faith and stole from the others, and men like Elmer, who sought to disrupt work and to set us against each other and even a neighbor or two, cantankerous and backward, who sought to obstruct the things we were trying to do. And there were the tragedies of death in the barns and the stables—the death of a calf or a colt, not only valuable, but beloved by all of us. And most discouraging of all were the checks and obstacles growing out of the war when materials and labor and farm machinery were unavailable and the progress both of experiments and The Plan were set back from five to ten years. It was a discouragement shared by farmers over the whole of the nation. But we learned many things out of the trials and disappointments, as every farmer must learn.

During two summer years of the war we were immeasurably helped by the boys who came to us from cities to work on the farm. They were only fifteen or sixteen years old but they worked like grown men, knocking off work now and then to go swimming in the pools of the Clear Fork and Switzer's Creek. They were full of intelligence and enthusiasm and sometimes, when there was hay down and the barometer fell, they worked long hours, the long hours which every farmer must work when the weather is threatening and crops or livestock must be cared for. They went to town on Saturday nights and did chores on Sunday and never complained. Indeed, they liked the life well enough to return each year to spend the Christmas holidays working on the farm.

The whole field of agriculture represents perhaps the most fascinating of all callings because no program is ever finished and each goal attained can be held only by vigilance and intelligence and because it embraces all of science and philosophy and life. In no life, certainly is there so little monotony, in no life so much of richness for those with understanding. Each farm is a tiny world in itself, with each day its small play of tragedy, of comedy, of farce. Each day is in itself a cycle of the history of the earth.

This much we have done with the land that has come into our hands at Malabar. Each day the forest grows greener and thicker. Each year the soil grows darker and deeper and the crops a little heavier. Each year water in the brooks and streams grows more steadfast and clear. The fish and wild game increase in number as the fertility increases. No longer does the water after heavy rains rush across bare land below us. No longer does the soil vanish by the ton after each rain to darken the streams and leave our own field bare and sterile. On the thousand acres of Malabar, no living gully, however small, exists like an open wound today. Each year more water gushes from the springs to water the cattle and the sheep and feed the little brooks where the water cress, which tolerates only clear, pure, cold water, grows on the gravelly bed.

Where there was once little, we have abundance. The trees we planted bear pears and peaches, grapes and plums and apples. The grapes trained along the fences that border the gardens hang green and yellow and purple in September. The gardens grow beans and cantaloupe, watermelon and broccoli, peas and carrots and sweet corn and asparagus and a score of other vegetables, not only for the season but for frozen storage all through the long winter. There are fat cattle and lambs and pigs in the fields and eggs and chickens and milk and cream and butter. The bees work for us and the trees provide maple syrup. Geese, turkeys, guinea fowl, ducks and fighting chickens wander at large over the fields, woods and ponds. Everywhere there is wild game, quail and pheasant, raccoon and possum, rabbit and squirrel and they too find abundance in the enrichened fields, in the orchards and the walnut and hickory and beechnut of the woods and the wild berries and hazelnuts of the fence rows.

The law of Nature is not that of scarcity but of abundance, and we have followed that law as nearly as we understand it. In all the period of shortages and rationing, we at Malabar took little or nothing from the needed corner stores of the nation and we contributed much. In our small world, since the very beginning, we have had both self-sufficiency and abundance. It was there to be had; to achieve it only required application of energy and intelligence. We have lived well at Malabar and the record is there in the fat, heavy cattle, in the brightness and health of every child on the place. We have been beholden to no one in what we have accomplished. In good times we have done well and in bad times we have always had security and shelter and food and above all else the dignity and self-respect without which life becomes for the reasonable man, unendurable.

I think we have reason to be proud of our record, for we rebuilt the very soil while we worked it, gaining a little way toward our ends even while we produced our crops and our animals. On the whole, we have lived together in harmony and decency with mutual self-respect and co-operation and as much privacy as we desired; and all that is not easy for as many as thirty-five men, women and children existing on a small co-operative democracy of a thousand acres. I think the record must imply a considerable degree of abundance and health and contentment. We have lived in peace with our neighbors and helped them and received help from them, I think largely because we were engaged upon a project of considerable size which required a concentration of energies that led us to mind our own business.

We have been rewarded, not only in terms of material plenty, but in countless other ways in terms of health and the spirit, and we have learned much out of the very soil itself. For the children the rewards have been greater possibly than for the adults. There has been health and good food and fields and woods to roam over, animals to care for, streams to fish and swim in, and all those contacts with air and earth and water which make for wisdom and understanding and judgment and for those resources later in life which are indestructible and far beyond either fame or riches in the long and trying span of life. They have learned, I think too, the great importance and solace of work, not the aimless, monotonous work of riveting and fitting together nuts and bolts, but of work which creates something, work which is richly its own reward, within the natural scheme of man's existence—the kind of work which contributes to progress and welfare of mankind and the plenty of the earth upon which he lives. Wherever they go in afterlife they will possess the knowledge of the fields and the brilliant beauty of a cock pheasant soaring above the green of a meadow in October. They will know how things grow and why. They will understand what goes on above the earth and in it. If they are ever bored and defeated it will be in the narrow streets of the city or in some dark office or the thunderous shed of a factory. They will, I think, understand what is decent and tolerant in life and comprehend both the evils of selfish exploitation and the evils of a regimented world in which human dignity and the soaring quality of the human spirit are cramped and stifled.

I believe that one day our soil and our forests from one end of the country to the other will be well managed and our supplies of water will be abundant and clean. I believe that there will be abundance for all as God and Nature intended, an abundance properly distributed when man has the wisdom to understand and solve such things. I believe there will be no more floods to destroy the things man has worked to create and even man himself. I believe that the abomination of great industrial cities will become a thing of the past, that men and women, and above all else, the children, will live in smaller communities in which there can be health and decency and human dignity, and that when that time comes, the people then living will look back upon us and the stupidity of our times as we look back with unbelief at the squalor and oppression and misery of the Middle Ages.

I am not a partisan of those who believe that in this country we have passed the zenith of our wealth and well-being. The possibilities of the future are boundless. Until now, we have destroyed as much as we have constructed and we are beginning at last to pay for that destruction in terms of reduced living standards and health and intelligence, in subsidies and all manner of politico-social-economic short cuts, tricks and panaceas. There is a tremendous job ahead of reconstruction and restoration, a job quite as big and infinitely more complicated than the job of subduing the wilderness by the first settlers. What we need is a new courage and a new race of pioneers, as sturdy as the original pioneers, but wiser than they—a race of pioneers concerned with the physical, economic and social paradise which this great country could be, if there were fewer exploiters, fewer selfish minorities, fewer self-seeking, vainglorious politicians, fewer social and economic panaceas and fanatics. These new pioneers will have to be men who sit not in libraries working out theories, but men who understand the people of this country and the illimitable wealth of its natural resources and beauties, and above all the fact that there is wealth for all and a good life and that it is founded, as is the wealth and well-being of every sound nation, upon its soil, its water and its forest. When there is no more soil, there will be no more nation, and the American civilization, even in its crude materialistic and mechanical manifestations, will wither and pass away.

For myself, I am deeply grateful to my parents and my grandparents for the life they gave me as a child in this rich Ohio country, for with it came the resources in life which are indestructible. They are bulwarks against fate, against wealth, against ambition, against poverty, against defeat. An acre is as good an anchor as fifty thousand, for in that acre, as Fabre well knew, there is the whole of a universe and the answer to most of man's problems.

And I am deeply grateful to the French for what I learned from them of the earth, of human values and dignity and decency and reality. And I am grateful to Louis Gillet, dead now of a heart broken by the humiliation of France, for the long talk of the evening in the moonlit forest of Ermenonville while we listened to the calling of the amorous stags, for he sent me back to the county where I was born, to Pleasant Valley and the richest life I have ever known.

Pleasant Valley was written on trains, in hotels, on ships, in the busy farm office at Malabar. I think some of the living quality of a busy life and some of the quality of trees, brooks and animals got into the pages, as well as the personalities of the countless farmers, business men, agricultural professors, and county agents who had their part in this book.

—LOUIS BROMFIELD

May 6, 1945

Printed in the United States
by Baker & Taylor Publisher Services